Five More Golden Rules

Knots, Codes, Chaos,
and Other Great Theories of
20th-Century Mathematics

John L. Casti

John Wiley & Sons, Inc.

New York ■ Chichester ■ Weinheim
Brisbane ■ Singapore ■ Toronto

W9-CYG-672

Copyright © 2000 by John L. Casti. All rights reserved

Published by John Wiley & Sons, Inc.
Published simultaneously in Canada.

Library of Congress Cataloging-in-Publication Data:

Casti, J. L.
    Five more golden rules : knots, codes, chaos, and other great theories of
    20th-century mathematics / John L. Casti.
        p.   cm.
    Includes bibliographical references and index.
    ISBN 0-471-32233-4 (cloth : alk. paper)
    1. Mathematics Popular works.   I. Title.
    QA93.C383   2000
    510—dc21   99-39544

Printed in the United States of America
10 9 8 7 6 5 4 3 2 1

Contents

CHAPTER

1

The Alexander Polynomial

Knot Theory

Knot History, Mathematics

According to legend, one of the world's first knot theorists was the simple Greek peasant, Gordius, who was named king of Phrygia when he arrived in a public square in an oxen cart around 1200 B.C., thus fulfilling an oracle's pronouncement that the future king would arrive in a wagon. In founding the Anatolian city of Gordium, the peasant gave thanks to the gods by dedicating his oxcart to Zeus, and tying it with an unusual knot. Then, as now, oracles were in ample supply, and another soothsayer came forward to claim that whoever untied the knot would rule all of Asia (at that time, meaning Persia). History records that in 333 B.C., Alexander the Great reached Gordium, was shown Gordius's chariot, and immediately "untied" the intricate knot with one fell swoop of his sword, thus giving rise to the proverbial term "solving a problem by bold action." This ancient story also gives rise to the principal problem facing knot theorists today: How difficult is it to untie a knot?

As with a lot of modern mathematics, the first important study of knots is due to Gauss, who while investigating electromagnetism in the latter part of the nineteenth century, developed methods for calculating properties of knots that he identified with the linking of electric and magnetic fields. A bit later, Lord Kelvin proposed the notion that atoms could be classified according to different types of knots, which led mathematicians such as Tait and Kirkman to produce tables of knot types. These efforts underscore the crucial problem of determining when one knot is the same as another. In what follows, we shall focus mostly on this question, as it remains the most important question in knot theory and is still only partially solved.

Amusing as it is for historical reasons, the Gordian knot is pretty far from the thoughts of theorists today interested in developing methods for distinguishing one knot from another. Knot theorists are motivated by far more important—or at least intellectually more challenging—questions than who is to rule Persia. For example, the strands of DNA residing at the center of every cell of your body are tightly coiled and tangled when packed into genes. But to replicate itself in the process of cell division, a DNA strand must uncoil itself. This unknotting process is carried out by enzymes that serve as microscopic versions of Alexander's sword, slicing through the strands of DNA to unknot them, and then "gluing" the ends back together again. So being able to say how difficult it is to untie a knot can help in determining how fast these enzymes can carry out their activity—as well as help us develop entirely new enzymes that might do things better, faster, or both. Much the same story can be told about questions in polymer chemistry. Even in the rarefied heights of elementary particle physics, there is growing evidence that the constituents of matter itself are nothing more than ultramicroscopic strings of energy coiled up in particular configurations. We'll see these and other examples of ways that knot theory works in practice later in the chapter. But for now let's take a harder look at what mathematicians mean when they talk about a "knotty" problem.

Have Knot, Will Unravel

Intuitively, a knot K is simply a closed curve sitting in ordinary three-dimensional space. Unfortunately, this definition is not quite acceptable because it admits offbeat curves containing an infinite sequence of smaller and smaller knottings, as well as other undesirable pathologies. So it's technically better to define a knot as a simple, closed curve composed of a finite—but possibly very large—number of straight line segments. It can be shown that all the properties of continuous curves we want for studying knots can be recovered from this definition by letting the number of segments approach infinity. So when pictures in the chapter show knots as continuous curves, bear in mind the straight-line definition underlying the figures.

A lowly cockroach crawling along any knot will see it as being just a one-dimensional closed curve and will have no sense of the three-dimensional space in which the knot sits (technically, its embedding). So it is the way the curve sits in an embedding space that distinguishes one knot from another. But why must the curve be embedded in three-dimensional space, rather than the plane, or even in four-dimensional space? The answer is that the plane is too small to contain a knot because the places where the curve crosses over itself to form the knot would have to stick up out of the plane. On the other hand, four-dimensional space is too big because any knot can be unraveled there, given the extra degree of freedom. So just as with Goldilock's porridge, ordinary three-dimensional space is not too big and not too small, but just right as the embedding space within which to talk about interesting knots. The simplest possible knot, which is not really a knot at all, but the opposite of

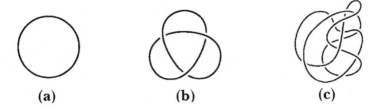

Figure 1.1. The unknot (a) and two versions of the trefoil knot (b, c).

a knot—the *unknot*—is shown in part (a) of Figure 1.1. Part (b) shows the simplest nontrivial knot, the trefoil or overhand knot. And just to show how complicated things can get, part (c) of the figure shows a nonstandard picture of the same trefoil knot. This illustrates nicely why there is a problem in trying to distinguish one knot from another, because it is far from obvious that the two pictures of the trefoil really represent exactly the same knot.

What exactly do we mean when we speak of one knot being equivalent, or "the same," as another? Topologists say that two geometric objects are "the same" if it is possible to continuously deform one of the objects by stretching, squeezing, twisting, turning—but not cutting or smoothing kinks—into the other. So, showing that a knot is nontrivial amounts to proving that it is not equivalent to the unknot. Speaking slightly more technically, determining the equivalence of two knots is a special case of the much more general "placement problem." An example is the rotation of a triangle drawn in the plane. Suppose we call the initial position of the triangle T_1 and then rotate the triangle by 90 degrees to a new position T_2. We then ask: Is there a transformation f that takes the position T_1 and smoothly moves it to position T_2? Here the answer is very simple. The requisite transformation is simply the operation that rotates objects in the plane through 90 degrees. In such situations, we would say that the triangles T_1 and T_2 are *equivalent* because they differ only by the transformation f.

Knot theory deals with a special case of the placement of a simple, closed curve in ordinary three-dimensional space. If we then ask if two such curves are equivalent, this means we want to know if there is a smooth transformation that takes one knot into another. If such a transformation exists we say that the two knots are equivalent; otherwise, they are essentially different.

All knots are closed curves, and hence are one-dimensional objects living in three-dimensional space. Thus, we can usefully study a knot's properties by looking at the shadow of the knot in the plane. In other words, we project the knot onto the plane, and then carefully distinguish over- and undercrossings. The diagram of the trefoil knot shown in Figure 1.1(b) is one such projection. Note, however, that there may be many different projections of a given knot (see Figure 1.2). So to effectively employ the knot diagram to study the question of whether one knot is equivalent to another, we need to identify properties of a diagram that remain unchanged for all possible projections of the knot. Let's make this point a bit more explicit by applying color to the knot.

What Color Is Your Knot?

One of the most famous conjectures in mathematics is the Four-Color Conjecture (FCC), which asserts that every nontrivial map drawn in the plane can be colored with no more than four colors. In 1976, Wolfgang Haken and Kenneth Appel produced a computer program that examined several thousand special configurations, concluding that the FCC is true. Because it is practically impossible for mathematicians to survey this "proof," the Haken-Appel result is still shrouded in doubt as to its acceptability as a bona fide piece of mathematics. Thus, the real importance of the computer solution of the FCC is not the verification of the conjecture, but rather the issues it opened up in the philosophy of mathematics.

Figure 1.2. Two projections of the trefoil knot.

The main philosophical issue that the Haken-Appel result opened up is about the nature of mathematical proof. What does it mean to say that we have "proved" a particular proposition? Prior to this computer-aided proof, mathematicians tacitly assumed that one of the things a "proof" meant is that every step of logical inference in the proof could be checked and verified. But with computer proofs such a verification procedure is effectively impossible because there may be billions or trillions of such steps and it would take far longer than the age of the universe for even an army of mathematicians to go through and check, one-by-one. In fact, the situation is even worse than this because it is known that every computer makes hardware mistakes at a known rate. So, for instance, the machine used by Haken and Appel can be guaranteed to, say, flip one bit in its memory every 100 million operations. Does this matter in the final result? Well, it depends on when *exactly* the bit is flipped and where it happens in the program the machine is running. If it happens when the machine is processing, say, a COMMENT statement, then it doesn't matter at all. But if at that critical moment the machine is deciding whether a particular quantity is positive or negative in order to decide what step in the program to branch to, a bit flip may send the program off onto an entirely wrong course. So is a proof really a proof when it cannot be checked and it is known to contain errors? That's the essence of the philosophical issue raised by the computer "proof" of the Four-Color Conjecture.

But the FCC is only the tip of an iceberg when it comes to coloring problems in mathematics. No such controversy surrounds the analogous question of coloring of knots. As we said, the principal goal of knot theorists is to discover ways to decide whether two knots are "the same." So a "colorful" approach to the construction of such decision procedures may be in order.

Suppose you try to unknot the trefoil. You twist and turn the strands, bend one over the other and, in general, do everything you can think of short of actually cutting the knot. But to no avail. Does this *prove* that the trefoil cannot be unknotted? Any self-respecting mathematician would say no, not at all—it simply shows that you are not clever enough to find a way to untie the knot. Let's look at a way to prove that the trefoil cannot be unknotted.

Our goal is to assign a number to each knot diagram in such a way that different diagrams of the same knot are assigned the same number. In that case, the number is called a *knot invariant* because the value of this number depends only on the knot and not on the particular diagram used to characterize it. If the value of this invariant is different for two separate diagrams, then the diagrams represent different (i.e., inequivalent) knots. Note, however, that it may happen that it does not work in the opposite direction: two different knots may have the same invariant, but if two knots have different invariants, the knots must also be different. We'll see examples of this later. Now let's construct one such invariant using the idea of colorability of a knot.

Consider all possible ways to color a knot diagram in three colors, so that each connected arc of the diagram is painted in one of three colors. Then at each crossing point the arcs coming together may carry one, two, or three colors. A coloring is said to be *proper* if there are no crossings where two colors come together. In other words, the arcs at each crossing either each have a different color or they all have the same color. Figure 1.3 illustrates this point with the Figure-8 knot.

We now have the following result:

COLOR INVARIANT THEOREM   *The number of different proper colorings of a knot diagram is a knot invariant.*   ∎

Proofs of this result are available in the material cited in the Notes and References section.

Now we are in a position to prove that the trefoil knot cannot be untied. Recall that untying a knot means that the knot is equivalent to the unknot. So what we need to show is that the trefoil cannot be transformed to the unknot by any smooth transformation of three-dimensional space.

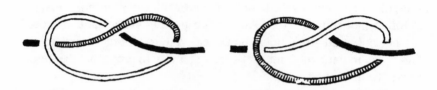

Figure 1.3.   Colorings of the Figure-8 knot.

Consider the proper coloring of the trefoil shown in Figure 1.4. The total number of such colorings is nine: six proper colorings using all three colors (one arc can be painted in any of the three colors, leaving a choice of two colors for the second arc and one color for the last arc) plus three monochromatic colorings in which all arcs are painted the same. But the unknot has only three proper colorings, all monochromatic. Thus, the coloring numbers of the trefoil and the unknot are different, and hence, the two knots must be different. In particular, this shows that the trefoil cannot be untied because untying it means transforming it into the unknot, which we now see is impossible.

Unfortunately, the coloring number is not a very powerful invariant because it doesn't even allow us to prove that the Figure-8 knot (Figure 1.3) cannot be untied, as each of its three proper colorings is monochromatic. So we will have to continue our search for stronger knot invariants.

The most apparent property of knot diagrams such as those of the trefoil in Figures 1.1(b) and 1.1(c) concerns the places where the strands of the knot cross over each other. Such crossings are technically termed *doublepoints*. Examination of these two diagrams of the trefoil shows immediately that the number of such crossings depends on the particular projection of the knot. However, it can be shown that the smallest number of crossings for a given knot is independent of the projection. The smallest number of crossings is called the *crossing number* of the knot. So, for instance, the trefoil has the crossing number 3, whereas the unknot in Figure 1.1(a), with no crossings, has crossing number 0.

Figure 1.4. A proper coloring of the trefoil knot.

The crossing number is thus an invariant of the knot, in the sense that it depends only on the intrinsic "knottiness" of the closed curve and not on the way it happens to be projected onto the plane. Therefore, we can use it as a crude filter to separate some knots from others. For instance, the trefoil with crossing number 3 cannot possibly be deformed into the Figure-8 knot shown in Figure 1.5, which has crossing number 4. Unfortunately, the crossing number is a pretty crude invariant because there are *lots* of inequivalent knots having the same crossing number. For example, it has been shown that there are exactly 9,988 different knots having at most 13 crossings. Consequently, we need much more subtle invariants if we hope to be able to separate this set of nearly ten thousand knots into distinct classes.

To illustrate the concept of the crossing number, Figure 1.6 shows a tabulation of all inequivalent knots have eight or fewer crossings. It is of some aesthetic interest to note that the famed German Renaissance painter Albrecht Dürer created six ornamental knots stimulated by knot designs of Leonardo da Vinci. These six Dürer knots are shown in Figure 1.7. But we still need better ways than the crossing number to separate knots.

A knot is a simple, closed curve, so it makes sense to think about starting at some point on the knot, and then traversing it and returning to your starting point. The direction one moves in such a traversal of the knot is called the knot's *orientation*. This is a basic aspect of a knot that is of great value in finding invariants. We can also attach a sign to each doublepoint of a knot. We do this by using the French traffic rule that cars on the right approaching an intersection have right-of-way. This means that a doublepoint is "positive" if the strand passing under the

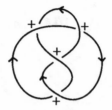

Figure 1.5. The oriented Figure-8 knot.

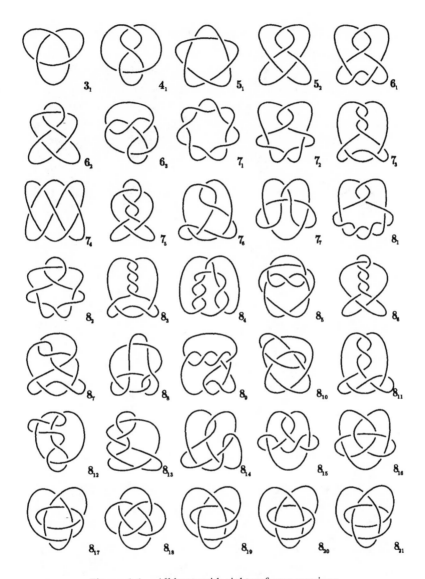

Figure 1.6. All knots with eight or fewer crossings.

doublepoint has the right-of-way, whereas it's "negative" if the strand passing over the doublepoint has priority.

Obviously, a given knot can have two separate orientations, each of which leads to attaching a sign, plus or minus, to the crossing number of the knot. Figure 1.5 shows the Figure-8 knot oriented such that each

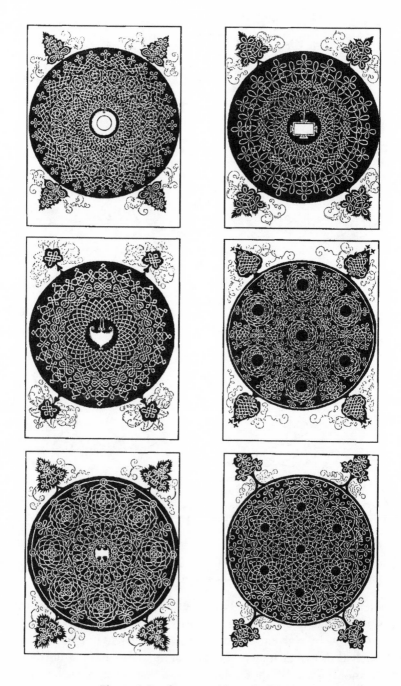

Figure 1.7. Ornamental knots by Dürer.

doublepoint is positive, so that its signed crossing number is +4. Taking the mirror image of the knot diagram by interchanging under- and over-crossings, the crossing number of the knot changes sign. We'll come back to orientation in a moment. For now, let's return to the question of unraveling a knot.

If we are given two knot diagrams of the same knot, for example, the diagrams of Figures 1.1(b) and 1.1(c) for the trefoil, we should be able to rearrange one of the diagrams to become the other. Thinking of the knot as being made out of a thin, flexible piece of string, such a rearrangement amounts to deforming the string in space in certain ways. Note, however, that doing something such as shrinking a part of the knot to a point in order to make the knot disappear (that is, become the un-knot) is not allowed, just as it's not possible to get rid of a knot in a string by drawing the string tighter and tighter. In 1926, the German mathematician Kurt Reidemeister proved that a given knot was equivalent to another if and only if the knot diagram of the first could be transformed into that of the second by a sequence of what are now called *Reidemeister moves*. He identified three such moves, which are shown in Figure 1.8.

A type I move allows us to put in or take out a twist in the knot. The second Reidemeister move, type II, permits the addition or removal

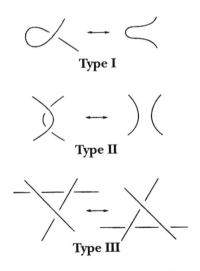

Type I

Type II

Type III

Figure 1.8. The three types of Reidemeister moves.

of two crossings, and the type III move allows us to slide a strand of the knot from one side of a crossing to the other side. These moves can be thought of as the legal moves that can be used in changing a knot diagram; they do not involve any sort of Alexandrian "cheating" by such things as cutting.

An interesting use of the Reidemeister moves is displayed in Figure 1.9, where a lengthy sequence of moves shows that the Figure-8 knot at the upper left of the first row is actually equivalent to its mirror image knot at the lower right. Knots such as this that are equivalent to their mirror image knot are called *achiral* by chemists, and are important in the analysis of some chemical compounds.

Before continuing the discussion of how to distinguish one knot from another, let's pause and look at the way even such simple ideas as the crossing number and the Reidemeister moves allow us to shed light on processes going on inside living cells. Along the way, we'll look at an another invariant of a knot, one related to the difficulty of untying it.

Unwinding DNA

Consider a single cell from a living organism. In most higher animals, all such cells are eukaryotic, which simply means that they have a nucleus. Inside this nucleus is a lot of cellular machinery used to create copies of the cell from the instructions coded into the cellular DNA, which itself is also wrapped up in the nucleus in the now famous double helical geometry. We can regard most DNA as a double strand, consisting of four bases: A = adenine, T = thymine, C = cytosine, and G = guanine.

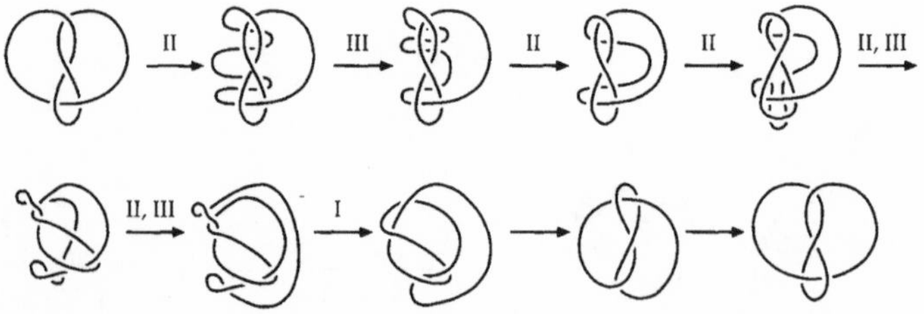

Figure 1.9. Reidemeister moves showing the achirality of the Figure-8 knot.

A ladder is formed by bonding between the base pairs, with A bonding to T, and G bonding to C. The base-pair sequence for a segment of double-stranded DNA is obtained by reading along one of the two backbones. Such a sequence is a word in the language of biology, formed from the alphabet {A, G, C, T}. In the Watson-Crick double helix model of DNA, shown below in Figure 1.10, the ladder is twisted in a right-handed helical pattern, with an average of about $10\frac{1}{2}$ base pairs per full rotation around the helix (this is called the *pitch* of the strand).

In the cellular nucleus, the DNA molecules are intertwined millions of times, as well as tied into knots and several orders of coiling to fit within the nucleus. Now, suppose we enlarge the cell's nucleus to the size of a basketball. On that scale, the cellular DNA would be more than 120 miles of tightly coiled material the thickness of thin fishing line. When it comes time for the cell to build a copy of itself, this bundle of nucleotide bases must be unknotted. How it does this is an extremely important practical problem, because the unknotting must take place rather quickly for replication, recombination, and cellular transcription to occur. In effect, the cell has to solve a topological problem.

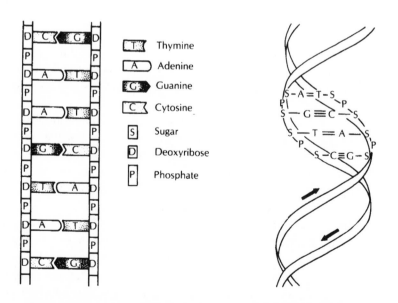

Figure 1.10. The double helix structure and base-pairing rule of DNA.

15

Compounds called *topoisomerases* are what perform this unknotting of cellular DNA, whereas other enzymes, called recombinases, break the DNA apart and recombine the ends by exchanging them. These compounds follow rules very similar to the kind of rules outlined by Reidemeister for unknotting knot diagrams. Let's see briefly how this works.

In Figure 1.11 we see three electron micrographs showing two trefoil knots in the top section, and two other DNA knots of six crossings each. The knots of real double-stranded DNA in the micrographs are redrawn as knot diagrams in the right-side of the figure. In knot-theoretic terms, unknotting such a DNA strand amounts to transforming its knot diagram into that of the unknot by a sequence of Reidemeister moves. Nature faces the same problem, and uses what are called type II topoisomerases to change the sign of a doublepoint of double-stranded DNA (type I topoisomerases do the same thing for single-stranded DNA from + to −, or vice versa). Physically, this amounts to cutting the DNA and reconnecting it so that the strand that previously was on top of the doublepoint now passes under it. The minimal number of such cuttings and gluings that must be done to transform the knot to the unknot is called the *unknotting number* of the knot. This number represents a crude measure of the complexity of the knot.

It's easy to see that changing the sign of one of the crossings in the trefoil at the top part of Figure 1.11 changes the trefoil to the unknot. For instance, if the crossing at the top is reversed so that the right-hand strand goes under the crossing, the whole knot dissolves into a straight piece of string—the unknot. So the unknotting number for the trefoil is 1. It's rather less clear how many sign changes at doublepoints are needed to unravel the two six-crossing knots in the remainder of the figure. Unfortunately, in general, it is very difficult to compute the unknotting number. For example, you might think that it could be computed by taking any diagram and finding the smallest combination of crossings that must be changed to untie the knot. But this is not the case, as can be seen from the knot diagrams in Figure 1.12. The left part shows a diagram that cannot be untied by changing fewer than three crossings. But the right part shows the same knot in a different diagram that can now be untied by changing just the two crossings indicated. The moral here is that to find the fewest crossings to change, you may have to take a simple-looking diagram and redraw it to look more complicated.

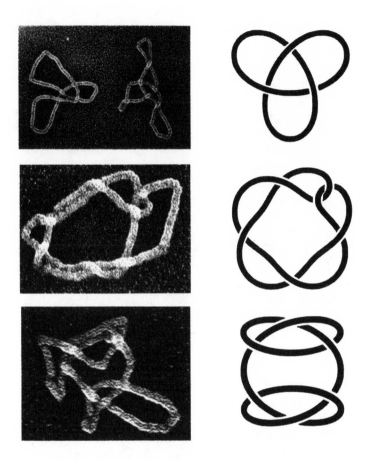

Figure 1.11. DNA knots and their knot diagrams.

There is one class of knots, however, for which we have complete information on the unknotting number. These are the so-called *torus knots,* which are closed curves drawn on the surface of a torus, the mathematical term for a doughnut. To tie what's termed a *(p, q)-torus knot,* you simply wrap a string *p* times through the hole in a doughnut, and then stretch the string so that it goes *q* times around the doughnut itself before tying the ends of the string together. (*Note:* For the knot to be tied on the torus without intersecting itself, the integers *p* and *q* cannot have any common divisor greater than 1; in other words, they must be relatively prime.) An example is the (4, 3)-torus knot shown in Figure 1.13.

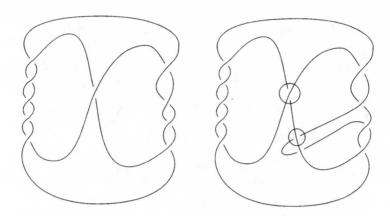

Figure 1.12. Two diagrams of the same knot requiring different crossings to untie.

More than 40 years ago, John Milnor of Princeton University conjectured that the unknotting number for a (p, q)-torus knot is the number $(p - 1)(q - 1)/2$. Recall that this is the number of sign reversals in doublepoints needed to transform the knot to the unknot. Given the difficulty in calculating the unknotting number, in general, this formula—if it's true—is nothing short of miraculous. Recently, Peter Kronheimer of the University of Oxford and Tomasz Mrowka of Caltech proved the Milnor conjecture as an unexpected spinoff from their work on four-dimensional manifolds. Even more recently, Lee Rudolph of Clark University has shown that the work of Kronheimer and Mrowka also proves a generalization of Milnor's conjecture that gives us a bound on how small the unknotting number can be for *all* knots. This result, known as the Bennequin conjecture, is discussed in detail in the references for this chapter cited at the end of the book.

Returning to DNA, the unknotting number gives a lower bound for the number of times a topoisomerase must act to convert knotted DNA to its unknotted form. As a result, it gives valuable information on determining reaction rates for topoisomerases. As DeWitt Summers, a knot theorist at Florida State University, expresses it, "If you have really complicated products that have a large unknotting number, it's going to take the enzyme a while to produce those." We'll see many more details of how knot theory enters into the behavior of DNA later in this chapter. But even this very sketchy outline shows the central role of unknotting in the way cells carry out their daily business. Now let's return to the search

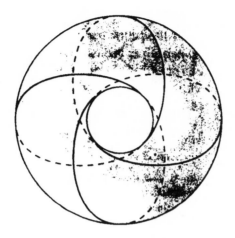

Figure 1.13. A (4, 3)-torus knot.

for finer-grained knot invariants because the quantities we've looked at so far are simply too crude to effectively separate very many knots.

Alexander's Great Invariant

When one thinks of the faculty appointments that started off the Princeton Institute for Advanced Study (IAS), all the usual suspects come to mind: Albert Einstein, John von Neumann, Hermann Weyl, Oswald Veblen, to name a few. The one name that's often overlooked, though, is that of James W. Alexander, who along with Einstein, Veblen, and von Neumann was part of the *original* IAS mathematics faculty members. But you don't get an appointment to a place like the IAS without having something going for you, and even a casual inspection of Alexander's work shows that he certainly belonged in such illustrious company.

A good-natured, outgoing gentleman who loved mountaineering, Alexander was the son of American artist, John Alexander, who painted murals in the U.S. Library of Congress. He was also one of the first of the new breed of topologists who sprung up in the Princeton University mathematics department in the 1920s, and prior to his departure across town to the IAS in 1933, Alexander introduced the idea of associating a polynomial (a mathematical gadget such as $t^3 + 5t^2 - 3$, which involves an indeterminate variable such as t raised to various powers, each power multiplied by a number) with a knot in such a way that the polynomial would serve as a powerful knot invariant. To develop the Alexander polynomial in full generality would involve a few more mathematical pyrotechnics than we'd care to go into in a book of this type, so let's look at how to find the Alexander polynomial of our old friend, the trefoil knot, to give the idea of how this invariant is determined in general.

First, pick a labeling of the crossing points c_i and the regions r_j of the trefoil knot diagram, with $i = 1, 2, 3$ and $j = 0, 1, 2, 3, 4$. Next, choose an orientation for traversing the diagram. Now place a dot in each of the regions that lie to your left as you pass *under* a crossing point. One such labeling of the diagram for the trefoil is shown in Figure 1.14.

Consider the crossing point c_1. With it we associate the linear relation in the indeterminate variable x:

$$c_1 = xr_2 - xr_0 + r_3 - r_4.$$

Here the index 2 denotes the dotted region you move into as you pass through the crossing point c_1, 0 denotes the dotted region you begin in as you move through the crossing point, 3 denotes the undotted region sitting to your right as you move into the crossing point, and 4 denotes the region on your right after you've passed through the crossing point.

In general, for crossing point c_i the appropriate linear relation is

$$c_i = xr_j - xr_k + r_l - r_m.$$

Here j is the index for the dotted region you move *into* as you pass through the crossing point, k is for the dotted region you *start* in as you

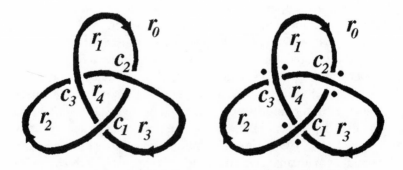

Figure 1.14. Labeling the trefoil diagram for computing the Alexander polynomial.

move through the crossing point, l is for the undotted region to your right as you move into the crossing point, and m is for the region to your right after you've passed through the crossing. Following this convention, for the three crossing points in Figure 1.12 we have the following relations:

$$c_1 = xr_2 - xr_0 + r_3 - r_4,$$
$$c_2 = xr_3 - xr_0 + r_1 - r_4,$$
$$c_3 = xr_1 - xr_0 + r_2 - r_4.$$

Using these relations, we form the following table whose columns are labeled by the different regions r_0 through r_4.

$$M = \begin{pmatrix} r_0 & r_1 & r_2 & r_3 & r_4 \\ -x & 0 & x & 1 & -1 \\ -x & 1 & 0 & x & -1 \\ -x & x & 1 & 0 & -1 \end{pmatrix}$$

In this array the entry appearing in row i, column j corresponds to the coefficient of the term r_j in relation c_i. For example, the first row is formed from the coefficients of the linear relation $c_1 = -x \times r_0 + 0 \times r_1 + x \times r_2 + 1 \times r_3 - 1 \times r_4$.

Now take any one of the crossing points and remove the columns from the array M that correspond to the two regions that are dotted at the chosen crossing point. So, for example, if we choose crossing point c_1 we would remove columns r_0 and r_2 from the array, leaving the reduced array

$$M' = \begin{pmatrix} r_1 & r_3 & r_4 \\ 0 & 1 & -1 \\ 1 & x & -1 \\ x & 0 & -1 \end{pmatrix}$$

Now for the coup de grace. We compute the determinant of the array M' (denoted $\Delta(M')$), to obtain

$$\Delta(M') = x^2 - x + 1.$$

This is the Alexander polynomial of the trefoil knot. Note, though, that the Alexander polynomial is not a knot invariant because it depends on how the crossing points and regions in the knot diagram are labeled, and there is no intrinsic reason why, for instance, what we've called region r_1 could not have been called region r_2 instead. But in 1928

Alexander proved the following remarkable result, showing that this choice of labeling is basically irrelevant:

ALEXANDER'S POLYNOMIAL INVARIANT THEOREM

1. *If the Alexander polynomial for a knot is computed using two differ- ent diagrams and labelings for the knot, then the two polynomials will differ by a multiple of $\pm x^k$ for some integer k.*
2. *The Alexander polynomial is invariant with respect to the Reide- meister moves. Thus, two knots differing only by a Reidemeister move must be equivalent.*
3. *A knot and its mirror image have the same Alexander polynomial. So a knot and its reflection are equivalent.*
4. *The Alexander polynomial is independent of the choice of how the knot is oriented.* ∎

To illustrate, we saw that the choice of labeling in Figure 1.14 led to the Alexander polynomial $x^2 - x + 1$ for the trefoil. But a different choice might have yielded the Alexander polynomial $-x^4 + x^3 - x^2$, which differs from the first by a factor of $-x^2$. Thus, by part (1) of the theorem, the two polynomials represent the same knot. For fun, the reader might like to try finding the Alexander polynomials for the Figure-8 knot of Figure 1.4 (answer: $x^2 - 3x + 1$).

Just how good an invariant is the Alexander polynomial? Does it enable us to distinguish a large class of knots? Or is it weak, like the crossing and coloring numbers that we have already examined? The answer is mixed. On the one hand, the Alexander polynomial is a *lot* stronger than these other invariants; but it's not perfect. It has been shown that all knots with less than nine crossings have a unique polynomial. So for such knots the Alexander polynomial is a complete invariant. That is, it distinguishes every such knot from any other knot with eight or less crossings. Unfortunately, knots with more than eight crossings have been found that have the same Alexander polynomial—but are known to be distinct knots. For example, the pretzel knot with 15 crossings shown in Figure 1.15 has the Alexander polynomial 1—exactly the same as the unknot. So the Alexander polynomial cannot distinguish between these two *very* different knots. Another failure of the Alexander polynomial is

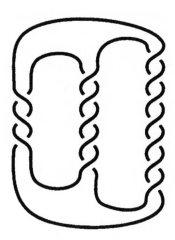

Figure 1.15. A pretzel knot with 15 crossings.

that it doesn't necessarily distinguish between a left- and a right-handed version of the same knot, such as the two versions of the trefoil shown in Figure 1.16.

Despite the fact that it is not a complete invariant, for nearly 60 years the Alexander polynomial remained the most powerful of the knot invariants. Then in 1984, Vaughn Jones of the University of California developed a new, even more powerful polynomial invariant. The

Figure 1.16. Left- and right-handed trefoils.

strangest part of Jones's remarkable result is that he wasn't even working in the area of knot theory at the time he made his discovery! Instead, Jones was studying the relationships between certain types of von Neumann algebras arising in quantum theory and the mathematical theory of braids. Moreover, at about the same time, another group of mathematicians working independently discovered further polynomial invariants that subsume both the Jones polynomial and the Alexander polynomial. This new polynomial involves two variables, x and y, which can be raised to negative as well as positive powers. This is in contrast to the Alexander polynomial, which has only positive powers of the single variable x. Just to see how these new polynomials look, the Jones polynomial invariant for the right-handed trefoil is

$$-x^{-4} + x^{-2}y + x^{-2}y^{-1}$$

whereas the polynomial for the left-handed trefoil is

$$-x^4 + x^2y + x^2y^{-1}.$$

Since these are not the same polynomials, we conclude that the left- and right-handed trefoils of Figure 1.16 are indeed not the same knot. Unfortunately, although the mathematical machinery needed to compute these new polynomials is straightforward enough to be programmable for a computer, it is a bit technical and too lengthy for us to present here. We note, however, that the computational time needed to compute the polynomial increases exponentially with the number of crossings in the knot. So this makes a knot with, say, 30 or 40 crossings impossible to check by computer. It's an open question as to whether a faster algorithm exists. The interested reader should consult the chapter references for more information on this computation.

Sad to say, even these vastly more powerful polynomial invariants *still* don't constitute a complete set of invariants. There are knots that are demonstrably inequivalent just by looking at their crossing numbers that still have the same Jones polynomial. For example, the two knots shown in Figure 1.17, the first of which has 8 crossings whereas the second has 10, share the polynomial

$$(-x^{-4} - x^{-2} + 2 + x^2) + (x^{-4} + 2x^{-2} - 2 - x^2)y^2 + (-x^{-2} + 1)y^4.$$

What does this polynomial really *mean* insofar as the trefoil knot is concerned? Or, more generally, what intuitive interpretation can we give

24

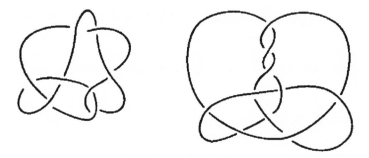

Figure 1.17. An 8- and a 10-crossing knot that have the same Jones polynomial.

to the polynomial in terms of the crossings, twists, and turns of a given knot? The answer is that no one really knows. Joan Birman, a leading knot theorist at Columbia University, describes the situation as follows: "There's a procedure for computing this polynomial, and there are all these different proofs that it really is attached to a knot type, but nobody understands geometrically what it means. ... it seems quite amazing that any one polynomial should detect so many different things."

As a halfway house between the pure theory of knots and knot invariants and their application in the real world, people have recently been looking at how to define the "energy" of a knot as a way of characterizing both a mathematical knot and the knotted molecules that it may represent. This is a bit complicated and technical to delve into here, but is of enough interest and importance that the more mathematically inclined reader is referred to appendix 1 of this chapter for the details.

Getting Physical

Everything in nature except a vampire has a mirror image. In particular, the amino acids making up the proteins in living organisms as well as the nucleic acids forming the cellular DNA all come in both left- and right-handed forms. These two forms, although chemically identical in the sense of being formed from exactly the same atoms, have entirely different chemical actions as a result of their "twisting" in opposite directions in space. Interestingly, in observations of galactic clouds in space, as well as in experiments in earthly chemical laboratories, it seems that both forms arise naturally in more or less equal proportions. Yet *all* life forms on Earth use exclusively left-handed amino acids to form proteins and right-handed nucleic acids to form the genetic material. As a consequence of this puzzling fact, you would starve to death on a world where the steaks were all made out of right-handed proteins because your body chemistry would be unable to break these proteins down to extract their energy. Let's see what knot theory can tell us about how molecules twist and turn to form the handedness they display, either in nature or by the design of humans.

The question of whether or not a spatial graph (i.e., a knot formed by straight lines connected in space) is the same as its mirror image (i.e., is *achiral*) is of special interest to chemists in their attempts to synthesize new drugs and other biochemical compounds. We can think of a vertex in such a graph as being an atom of some molecule, and the edges of the graph as connecting two atoms having a common bond. In this way, the spatial graph represents a model of the structure of a molecule. Note, though, that even if a molecule can be topologically deformed to its mirror image, it may be impossible to bend and twist the actual molecule through space into its mirror image because the rigidity of the bonds linking its atoms won't allow it. So a given molecule may be topologically achiral but not "chemically achiral." However, a molecule that is topologically chiral must be chemically chiral.

Attempts to *create* a real-world molecule whose structure is a given spatial graph began in the early part of this century. Because it is easy to analyze the properties of a graph, if chemists can construct a real-world molecule having the same structure as a graph, then its properties should be similar—or so goes the theory, anyway. In the 1970s, chemist D. M. Walba succeeded in synthesizing a molecule whose geometric structure is that of a Möbius band. This molecule, denoted M_3 for its three molecular bonds, is shown in Figure 1.18(a); the more general version of the molecule, M_n, having n such bonds instead of just 3, is shown in Figure 1.18(b). In the figure, the elements a_i represent the atoms of the molecule, and the straight lines are the bonds linking the various atoms. Geometrically, one can see here that the bonds twist in space to form the famous one-sided Möbius structure.

From a chemical point of view, both M_n and its mirror image, M_n^*, are the same. So one might conjecture that M_n is a chiral molecule. Knot theory can help us resolve this conjecture. We have the following very important result:

MÖBIUS BAND THEOREM *The Möbius band molecule* M_n *is chiral for* n $\geq$ 4. *But if* n = 3, *there does not exist a continuous, one-to-one transformation of three-dimensional space that transforms* M_3 *to* M_3^* *and that maps each edge* $a_i a'_i$ *to its mirror image edge* $b_i b'_i$ *for* i = 1, 2, 3. ∎

The problem with the case of M_3 is that we need to distinguish between the rungs (the segments $a_1 - a'_1$, $a_2 - a'_2$, $a_3 - a'_3$) and edges (the segments $a_1 - a_2$ and $a_2 - a_3$) of the "ladder" constituting the Möbius molecule, in order to preserve the ordering of the edges. One way to visualize this is to color all the rungs red and all the edges blue. If we insist that red rungs go to red rungs and blue edges go to blue edges, then M_3 would also be chiral.

The results of Walba just discussed show that knotted molecules can be synthesized. So what about molecules in the form of various graphs? Every molecule can, of course, be represented by some graph. But usually the graph is a simple one in that it can be drawn in the plane with no crossings (a *planar* graph). One place to look for molecules that are not planar is in the family of metals. The root source of the strength and rigidity of metals is that they form huge three-dimensional lattices, each atom of which is bonded to all of its nearest neighbors.

In 1981, chemists Howard Simmons III and Leo Paquette independently constructed molecules represented by a nonplanar graph called K_5. Their construction is shown in Figure 1.19, where the white circles represent oxygen atoms, and the black circles represent carbon atoms. (The Simmons-Paquette molecule also had many hydrogen bonds that are not shown in the figure.) If all the edges were the

Figure 1.18. Synthesized molecules (a, b) in the shape of a Möbius band (center).

27

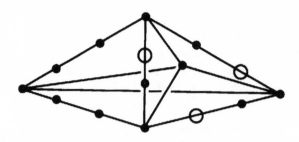

Figure 1.19. The Simmons-Paquette nonplanar molecule.

same in the K_5 molecule, it would be topologically achiral. However, the Simmons-Paquette K_5 molecule has edges of three different types: CH_2-CH_2 chains, CH_2O chains, and C-C single bonds. If one distinguishes these three types of edges by different colors, as was done for the Möbius molecule, it can be proved that this K_5 molecule is topologically chiral.

Let's now conclude our discussion of knots by returning to a more detailed examination of how they relate to the tangled molecular structures we encounter in biology.

All Tangled Up

In 1962 James Watson and Francis Crick of Cambridge University were awarded the Nobel Prize in medicine for their discovery of the celebrated double helix structure of the DNA molecule (see Figure 1.10). This discovery set the stage for the feverish work in molecular biology over the past several decades—work that has put us on the threshold of being able to actually write down the entire "program" for creating a human being.

As Figure 1.10 shows, a molecule of DNA can be thought of as two linear strands intertwined in the double helix geometry. The molecule can also take the form of a ring, thus opening up the possibility for it to become tangled or even knotted. Even more interesting is that the DNA molecule occasionally breaks, and while in the broken state it undergoes chemical changes. It then recombines into a new structure. Thinking of the molecule as a knot, this breakage and recombination is eerily reminiscent of the Reidemeister steps one goes through to untie a knot (except in knot theory we don't allow cutting or breaking the knot in transforming it to a simpler knot). So let's take a quick tour of this topological approach to the study of DNA.

A mathematical model for DNA is usually thought of as being a long, thin, narrow ribbon or band B, as in the two structures depicted in Figure 1.20. The two curves C_1 and C_2 forming the boundaries of the ribbon B represent the two strands of the molecule. If we pick a direction (an orientation) on the central curve C forming the axis of the strand, then a similar direction is induced on the curves forming the boundary of B. Since each of the closed curves C_1 and C_2 is a knot, we can think of the pair together as what knot theorists call a *link*. Intuitively, this is nothing more than two knots tangled together. So if instead of one string we take several and glue the extremities of each together, we have a link.

Once we have decided upon an orientation of the knot K, we can then define the *linking number*. Recall that an orientation gives a sign $+1$ or -1 to each crossing point in the knot diagram. If sign c_i denotes the sign of the ith crossing point, then the linking number of K is defined to be

$$\text{Lk}(K) = \frac{1}{2}\left[\text{sign } c_1 + \text{sign } c_2 + \cdots + \text{sign } c_n\right].$$

So $\text{Lk}(K)$ measures (roughly) how "tangled" the knot K is. The linking number is a knot invariant.

The linking number is very important for understanding the structure of DNA. For example, if we reduce the linking number of the double helix structure, the effect is to cause the molecule to twist and coil due to

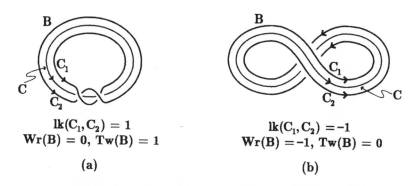

$$\text{lk}(C_1, C_2) = 1$$
$$\text{Wr}(B) = 0, \ \text{Tw}(B) = 1$$

(a)

$$\text{lk}(C_1, C_2) = -1$$
$$\text{Wr}(B) = -1, \ \text{Tw}(B) = 0$$

(b)

Figure 1.20. Two geometrical models of a DNA molecule.

freeing up some of the chemical bonds holding the molecule together. In nature, this can be done by having a topoisomerase act on the molecule.

There are other invariants noted as well in Figure 1.20. For example, the number of twists the DNA band B has along its axis is called the *twisting number,* denoted $\text{Tw}(B)$. And if we average the sum of the signs of the crossing points over all projections of the knot, then we obtain what is called the *writhing number,* $\text{Wr}(B)$, a measure of the degree to which knot B must bend out of a flat plane. Note, though, that these are invariants of the ribbon B thought of as a two-dimensional surface in space, not invariants of the one-dimensional curve characterizing the DNA molecule. These three invariants—the linking, twisting, and writhing numbers—are related by the following simple rule:

$$\text{Lk}(B) = \text{Tw}(B) + \text{Wr}(B).$$

The DNA molecule changes its linking number in the following fashion. First, a strand of DNA is cut at some location. Then a segment of DNA passes through this cut. Finally, the DNA reconnects itself. Interestingly, in the 1970s it was discovered that there is a single enzyme, topoisomerase, that causes both the cutting and the reconnecting of the DNA. The effect of such a topoisomerase action is either to move a piece of the DNA molecule to another position within itself or to add a foreign piece of DNA to the existing DNA strand. In either case, there is a genetic change.

Such a site-specific recombination is rather easy to illustrate. First, two points of the same or different DNA molecules are drawn together. The topoisomerase then goes to work, causing the DNA molecules to be cut open at two points on the parts that have been brought together. The loose ends are then recombined in a way that differs from the original molecule. Figure 1.21 shows a simple version of this process. Here the part from (a) to just before (b) shows the process of bringing the two parts together. This is called the writhing process. Then in part (b) the topoisomerase enzyme acts to form a complex, to which we can assign an orientation to the small part of the molecule on which the enzyme acts (the part inside the small circle in part (b)). Finally, the enzyme glues the pieces back together to form the compound knot (linked curves) in part (c). From here, the tools developed to study the properties of the knot can be invoked to analyze the structure of the molecule in greater detail, as well as to understand more deeply the process of DNA recombination.

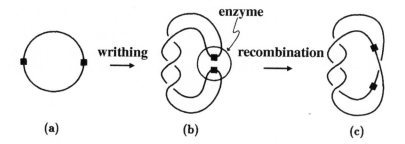

Figure 1.21. Site-specific recombination in the DNA molecule.

As this gets rather technical rather quickly, we refer the reader to the source material cited in the chapter references for details.

Appendix: Knots and Energy

Consider a knot K enclosed by a garden hose. Now turn on the water and let it flow through the hose, measuring the energy of the flow. This basic idea leads to what H. K. Moffatt termed the "energy spectrum" of the knot. This turns out to be another knot invariant, one that has very close connections to the final configurations of physical objects as different as DNA molecules, polymers, and plasmas. Given this disparate range of possible applications, it is worth looking just a bit harder at how this invariant is calculated—and what it might mean.

Basically, there are three steps in determining the energy spectrum of the knot K. First, we construct a knot field, which specifies the direction the water flows. This field is constrained to lie inside the hose enclosing the knot. The knot field has a well-defined energy, which leads to the second step. This step involves letting the knot and its associated field relax to a minimal-energy configuration. We accomplish this by letting the diameter of the hose gradually get larger and larger. It's necessary to do this expansion very carefully, however, to ensure that we don't end up with a configuration having zero energy. This could easily happen, for instance, if we simply scaled the field to zero. The expansion of the hose diameter must come to a halt when different parts of the hose come into contact with each other.

The third and final step in Moffatt's process is to recognize that there may be more than one minimal-energy configuration. This comes from the fact that the constructions start from a particular knot diagram,

and if the knot is sufficiently complicated it is possible (even probable) that starting from another diagram for the same knot will lead to a
different minimal-energy configuration. If we let m_i denote the energy
of the ith minimal-energy configuration, then Moffatt showed that the
minimal energy for this ith knot diagram will be

$$E_i = m_i \Phi^2 V^{-1/3},$$

where Φ is the flux of the fluid flow and V is the cross-sectional volume
element of the hose.

The quantity m_i is a dimensionless number, determined only by
the knot topology, which by virtue of its construction is a topological
invariant of the knot. The energy spectrum of the knot K is then the
sequence of numbers $0 \le m_0 \le m_1 \le m_2 \dots$. The actual minimum
energy, m_0, is the ground-state energy for the knot. Note, however, that
m_0 by itself is not enough to characterize the knot; one needs the entire
sequence.

To illustrate this idea of the energy spectrum, Figure 1.A1 shows
two different representations of the trefoil knot: the left side of each
representation shows the knot encased within the garden hose, and the
right side shows the inflation of the hose around the knot itself, which
is depicted by the dotted curve inside the hose. Here we also indicate
that the length L and area A of the hose is proportional to the one-third
and two-thirds power of the volume element, respectively. The top part
of Figure 1.A1 is the standard representation of the trefoil, for which
$m = m_0$. It is conjectured that the lower representation, which appears
more complicated geometrically, actually does have $m = m_1 > m_0$, but
so far there is no actual proof that this is indeed the case.

Using a different definition of the energy of a knot, Bryson, Freed-
man, He, and Wang proved some extremely interesting and important
results about crossing numbers and unknottings. In their work, the energy of a curve σ is defined to be

$$E(\sigma) = \int \int \left\{ \frac{1}{|\sigma(t) - \sigma(s)|} - \frac{1}{D(\sigma(t), \sigma(s))^2} \right\} |\dot{\sigma}(t)||\dot{\sigma}(s)| \, dt \, ds,$$

where $D(\sigma(t), \sigma(s))$ is the shortest distance between $\sigma(t)$ and $\sigma(s)$ on
the curve. Thinking of a knot K as being a closed curve in R^3, this
definition leads to the knot of minimal energy being a circle, that is, the
unknot. Moreover, the crossing number of a polygonal knot K of the

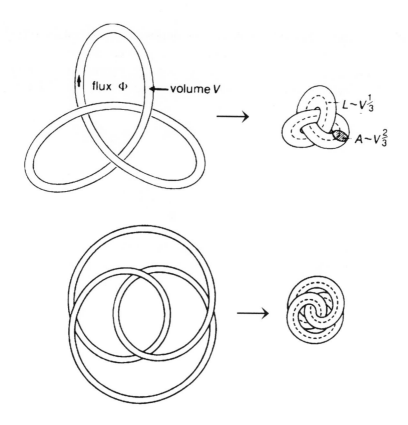

Figure 1.A1. Two representations of the trefoil knot encased in a hose.

sort we're considering in this chapter can be bounded from above by the quantity $(E(K) - 4)/2\pi$. In fact, the energy being finite is sufficient to ensure that the knot is of polygonal type. Finally, the number of distinct knots $K(n)$ having crossing number n can be bounded from above and below as follows:

$$2^n \leq K(n) \leq 2 \times 24^n.$$

This last inequality leads to the result that the number of distinct knots having energy less than or equal to M is approximately $(0.264)(1.658)^M$, which proves that only finitely many knot types occur below any finite energy threshold.

So we see that if knot K has a configuration of sufficiently small energy, then K is actually the unknot no matter how complicated it might appear. Intuitively, this suggests that it takes a lot of energy to make fluid flow around a truly complicated knot. In other words, a complicated knot is highly curved. The whole procedure can be reversed to shed light on the physical flow of fluids, as well. For example, the topology of knots may be employed to calculate lower bounds on the energy of fluid flows.

CHAPTER

2

The Hopf Bifurcation Theorem

Dynamical System Theory

What Is a Dynamical System?

Treasure Hunt is a game in which each player receives a list of clues that point to a prize. One version of the game involves having the first list of clues lead to a second list, which in turn leads to a third list and so on. The final list of clues then directs the player to the actual prize itself. The path followed by any one player in this type of game serves admirably as an example of a dynamical system; the mathematical apparatus of such a system enables us to study the appearance and disappearance of patterns and behaviors over the course of time. So let's follow the fortunes of a single player in a game of Treasure Hunt, calling our player John.

Assume for the sake of definiteness that the playing field for the treasure hunt consists of Central Park in Manhattan. The game begins with John receiving the initial set of clues while standing at, say, the Plaza Hotel entrance to the park, which we might conveniently term "ground zero." Decoding the clues on this initial list, John moves from the Plaza entrance to the foot of a large oak tree next to the zoo, where the second set of clues turns up buried beneath some leaves and branches. Contemplating these new clues, John soon ferrets out their meaning, and takes a hike to the opposite side of the park over by the lake. Digging into the sand by the shore of the lake, John quickly discovers the third set of clues, whose decoded message says to go to the delivery entrance behind the Tavern on the Green, where the "treasure," a coupon worth a dinner for two at any restaurant in the city, lies hidden beneath the third garbage can on the left. Figure 2.1 shows a pictorial account of John's tour through the park, moving from the starting point at the Plaza entrance to the promise of a fine meal. For later reference, the reader should note that Figure 2.1 also shows the path taken by Jim, one of John's competitors in the game, who started at the Fifth Avenue entrance to the park rather than at the Plaza Hotel. John's actions in following the clues from the Plaza entrance to the Tavern on the Green is a concrete example of what mathematicians have come to call a *dynamical system.*

There are three principal ingredients forming a dynamical system:

1. A playing field, or *space M,* on which the motion of the system takes place
2. A *rule v* stating where to go next from the system's current position in *M*

3. A *time set T* that describes the moments at which movement from one location in *M* to another takes place.

Technically, the space *M* on which the system's states "live" is called a *manifold,* and the rule of motion *v* telling you how to move from one point of *M* to another goes under the rubric of a *vector field.* So in the case of John and the treasure hunt, the manifold is Central Park, a rectangular region of Manhattan bounded on the north by 110th Street, on the south by 59th Street, on the west by Central Park West, and on the east by Fifth Avenue. The rule governing how John moves about on this manifold is simply the set of instructions saying where to go next, instructions coded into the lists of clues John found at the various locations in the park. So the three lists of decoded clues found by John constitute the vector field that determines a dynamical system that we might label "John." As for the time set *T*, if John were a kangaroo instead of a human being, hopping from one location to another instead of moving in a more or less continuous fashion would doubtlessly seem more natural. In that case, the dynamical system "John" would be called a *discrete-time* system because the movement from one state to another occurs only at the discrete set of moments *t* = 1, 2, 3, ... On the other hand, if motion takes place at every instant over some interval of real numbers, then we have a *continuous-time* dynamical system.

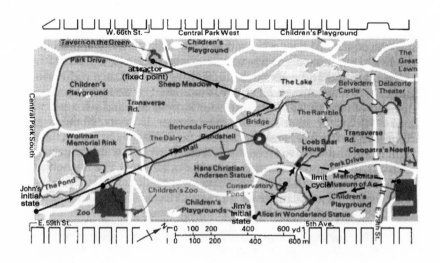

Figure 2.1. Treasure Hunt as a dynamical system.

The path John traces in moving from one location in Central Park to another is called the *trajectory,* or *flow,* of the system. Finally, John's starting point at the Plaza entrance is the dynamical system's *initial state.* It is clear from this setup, I think, that once the manifold M, the initial state, and the vector field v (the rule of state transition) are given, the system's trajectory is completely determined. To fix these ideas in more concrete terms, let's look at some simple examples of dynamical systems.

Insect Population Fluctuations

In 1976, the prestigious British scientific weekly *Nature* published an article by Robert May, a mathematical ecologist now at Oxford University, that served as one of the sparks touching off the explosion of current interest in chaotic dynamics. The theoretical centerpiece of May's path-breaking article turned out to be none other than a simple rule used to represent fluctuations in the population level of an insect colony. Here we consider a slightly simplified version of May's rule, termed the *logistic equation.*

Let's look at the logistic equation from an ecologist's point of view. First, we denote the insect population level at generation t by the symbol x_t. Also, to avoid having the population go shooting off to infinity, we scale the population so that levels beyond 1 signify an uncontrollable population explosion, whereas levels less than 0 denote extinction. In addition, we assume that the population is measured only at discrete times $t = 0, 1, 2, \ldots$

Now let's assume that if there is no competition for resources each adult produces α offspring and then dies. So the population in the next generation will be $\alpha \times x_t$. Under these conditions, the quantity $1 - x_t$ then represents a reaction to birth rates, describing the rate of death from effects due to overcrowding (e.g., competition for scarce resources). Consequently, if you have lots of crowding (x_t near 1), the population dies off quickly, whereas if there is little crowding (x_t near 0), overpopulation plays has little effect on new births. Under all these conditions and assumption, the rule of state transition for the insect population system becomes

$$x_{t+1} = \alpha x_t (1 - x_t), \quad t = 0, 1, 2, \ldots \qquad (*)$$

For this dynamical process describing the fluctuation in the number of insects, the manifold M is the unit interval, that is, $M = \{x_t : 0 \leq x_t \leq 1\}$, and the vector field $v(x)$ that describes how the system state moves about in this interval is just the rule $v(x) = \alpha x(1 - x)$. Finally, the time set T is discrete (but still countably infinite) and consists of the nonnegative integers $T = \{0, 1, 2, \dots\}$. So, as soon as the birth rate α and the initial population level x_0 are prescribed, the vector field $v(x)$ tells us how many insects will be present at any moment in the time set T. We'll see later how and why this simple-looking rule led to the "chaos revolution."

Radioactive Decay

It is known from laboratory experiments that the rate of decay in a sample of a radioactive material such as radium is directly proportional to the amount of matter present at any moment. So if $x(t)$ represents the number of grams of radium present at time t, then the manifold of states for the dynamical process describing radioactive decay is just the set of positive real numbers $M = \{x : x > 0\}$ (because the matter cannot ever disappear entirely). Now suppose the constant representing the rate of decay is denoted by k, which, of course, must be a positive real number. Its actual value must be determined experimentally. Then the vector field for this dynamical system is $v(x) = -kx$. So if there are x grams of radium present at time t, the change in the amount of radium present over a small time interval dt is simply $v(x) \, dt$. In elementary calculus notation this translates into the following *differential equation,* which describes how much radium is present at any moment of time:

$$\frac{dx}{dt} = -kx(t), \quad k > 0, \quad t \geq 0.$$

The time set T for this system consists of the set of nonnegative real numbers $T = \{t : 0 \leq t \leq \infty\}$. Again, once the rate of decay k and the initial amount of radium $x(0)$ are given, the vector field fixes the radium level for all future times. And, in fact, for this especially simple system we can write a closed formula for the state of the system at any time t as $x(t) = e^{-kt}x(0)$. Generally speaking, however, there is no such simple formula characterizing the state of a dynamical system. This is because the set of all possible dynamical systems is *much* larger than the set of elementary functions such as exp, sin, and cos that have turned out to be so useful that we have attached names to them. This fact has given rise to a whole branch of mathematics devoted to studying the qualitative character of the trajectory of a dynamical system. These qualitative properties include, for example, how the trajectory of the system "twists," what its long-run behavior looks like, and other characteristics of the system's behavior that don't change if you "speed up" the rate of motion along the trajectory. The concepts and techniques of this branch of modern mathematics are the focus of this chapter.

Good examples of dynamical systems are everyday processes such as the movement of the Dow Jones Industrial Average on the New York Stock Exchange, the movement of road traffic on the Pasadena Freeway, and the motion of the planet Jupiter as it goes along its path around the Sun. In each of these cases, there is a set of states (price movements, positions and velocities of cars, position and velocity of a planet), along with a rule of motion (how prices move from one level to another, how cars move from one location to another, and how Jupiter moves from one place to another). In the latter case, we actually know this law of motion, the vector field. In the first two cases, however, we don't. In fact, a large part of the scientific enterprise is devoted to just this task: finding the right rule of motion for some real-world dynamical system such as a stock market or a freeway.

Returning to the radioactive decay example, we can geometrically represent the motion on the state space M of the system by the diagram in Figure 2.2. Here we see the magnitude of the motion (the length of the lines marked v) becoming progressively smaller as we approach the origin. This picture suggests that there is something special about the system at the point where the vector field $v(x)$, the lines labeled v, vanishes. And so it is, as we'll see in just a moment. But first, a small digression.

The two examples of dynamical systems just given involve a state manifold that is continuous. Some readers might wonder why we can't

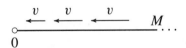

Figure 2.2. Vector field for radioactive decay.

have either a continuous or a discrete state space M, just as we had the option of using either a continuous or a discrete time set T. The short answer is that we can indeed have both types of state manifolds. When we discretize *both* the state manifold and the time set, the end result is a type of dynamical system called a *cellular automaton*. Although we won't make any use of this class of dynamical systems in this chapter, cellular automata have become increasingly popular in recent years as a way of representing phenomena as diverse as vertebrate skin patterns and the flow of fluids in a pipe. So for the sake of completeness let's just take a short look at what one of these cellular automata is like.

Growth of Plants

In the late 1960s, Dutch biologist Aristid Lindenmayer proposed the following type of cellular automaton to model the development of filamentous plants. The state manifold for such an *L-system* (so named in Lindenmayer's honor) consists of the countably infinite integer points, or "cells," on the real line. Thus, $M = \{\ldots, -2, -1, 0, 1, 2, \ldots\}$. We assume that each of these cells can take on the value either 0 or 1 at each instant of time. The value 1 means there is a living cell at that location at that moment; otherwise, the cell is empty. (*Note:* Here the word "cell" is being used in two entirely different ways: to indicate a location in the L-system's grid and as a cell in the usual biological sense. Context should make it clear which meaning is being employed in a given statement.) Lindenmayer's model contains the novel feature that the number of cells in the automaton is allowed to increase with time according to a recipe laid down by the system's rule for state change, its vector field. In this way the model "grows" in a manner mimicking the growth of a filamentous plant such as the blue-green algae *Anabaena*.

The simplest version of an L-system involves a process that starts at time $t = 0$ with a single active cell having value 1. The vector field prescribing how the value at cell i changes is determined by its current value, together with the value of the cell immediately to the left of cell i, that is, cell $i - 1$. To be specific, the complete state transition at cell i for this L-system is given below in Table 2.1.

The interesting feature of the rule in Table 2.1 is the possibility for cell division that occurs when $i = 1$ and $i - 1 = 0$, that is, when there is no cell immediately to the left of an existing cell at location i.

Table 2.1. State transition rule for an L-system.

time t	time $t + 1$
($i = 0$, $i - 1 = 0$)	$i \rightarrow 0$
($i = 0$, $i - 1 = 1$)	$i \rightarrow 1$
($i = 1$, $i - 1 = 0$)	$i \rightarrow 11$
($i = 1$, $i - 1 = 1$)	$i \rightarrow 0$

If we take an initial state such that the first cell at the left has value 0, and the cell next to it has value 1, then the first four state transitions are $01 \rightarrow 011 \rightarrow 0110 \rightarrow 011011$. Already we see here the growth of the initial "seed" at cell location 2 to several seeds at cells 2, 3, 5, and 6.

We must admit that this string of 0s and 1s doesn't look much like what anyone would even charitably call a "plant." So to bring out more clearly the connection between L-systems and the real world of plants, let's soup up our notation a bit by extending the symbol alphabet to include the new symbols [and]. And while we're at it, let's also extend the state transition rule so that the four symbols of the L-system always transform in the following manner:

$$0 \rightarrow 1[0]1[0]0, \qquad 1 \rightarrow 11, \qquad [\rightarrow [, \qquad] \rightarrow].$$

To see how this rule works, suppose we start with the string consisting of the single symbol 0. With the above transformation rules, the first three steps of this automaton lead to the sequence of states

$$0 \rightarrow 1[0]1[0]0 \rightarrow 11[1[0]1[0]0]11[1[0]1[0]0]1[0]1[0]0$$

This still doesn't look much like a plant. But we can convert this type of string into a tree-like structure by treating the symbols 0 and 1 as line segments while regarding the two brackets [and] as branch points.

One way to get something that looks faintly plantlike out of all this is to leave all the 1 segments bare, and place a leaf at the end of each 0 segment. So, for instance, if we have the string 1[0]1[0]0, its stem consists of the three symbols not in brackets. These are a 1 segment beneath another 1 segment, which in turn is topped off by a 0 segment. Two branches, each with a single 0 segment, sprout from this string. The first branch is attached above the first segment, whereas the second branch occurs after the second segment. With respect to the direction

43

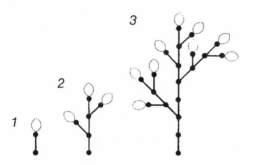

Figure 2.3. The first three generations of an L-system plant.

of the branch, the simplest convention is just to specify that for any given stem the branches shoot off alternately to the left and to the right. Figure 2.3 shows the first three generations of such a "plant" obtained by using the L-system grammar given above, together with the foregoing rule regarding leaves and branches.

No one claims that real plants grow in accordance with this primitive L-system's rules. The value of the L-system is that it shows how *some* plantlike features, such as branching and cell division, can be generated by very simple rules. This, in turn, suggests that although nature may manifest itself in what appear to be very complicated and intricate patterns, it may well be that extremely primitive rules underlie all this diversity and complexity.

With these fundamental ideas about manifolds, vector fields, and the like in our hip pocket, we can finally begin to tour the wonderland of dynamical systems. And there is no more appropriate place to start than to return to the curiosity noted earlier in the radioactive decay problem regarding the vanishing of the system's vector field as the motion in Figure 2.2 gets closer and closer to the origin.

In the Long Run

Without a doubt, the single most important thing we can ask about a dynamical system is what its behavior is like as time becomes large. What does the system do after all the short-term, transient behaviors have run their course? There are at least two reasons why this is the key question about which the theory of dynamical system revolves.

1. Many dynamical processes that we observe in real life, such as plant growth, radioactive decay, or a national economy, can be assumed to have been in operation for a long time. Consequently, any transient behavior associated with the settling-in time of such systems should have long ago washed out of the system, leaving the behavior we see as being that of the system operating in its so-called "steady-state" or "equilibrium" mode.

2. Think of a small cork bobbing about in the ocean near the periphery of a whirlpool. Clearly, even though it's not in the whirlpool itself, the cork moves in a spiral-like trajectory whose geometric nature is completely determined by the size, shape, and strength of the whirlpool. So it is also with dynamical systems; the nature of the trajectory determined by the system's vector field is organized by the nature of the system's "attractors"— the interior of the whirlpool, so to speak. These are the points in the state manifold that once entered can never be left—including as a special case those points at which the vector field vanishes completely (such as the center focal point of the whirlpool). The diagram in Figure 2.4 illustrates this point for the case of a system having a single attracting point C, whereas Figure 2.5 shows the same kind of diagram for a vector field whose long-run behavior is more like the circular motion of a whirlpool.

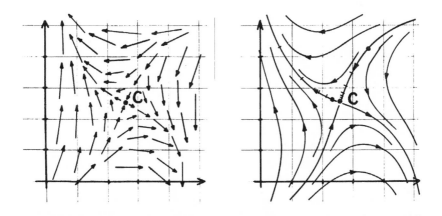

Figure 2.4. A vector field having a single point attractor C, and its trajectories.

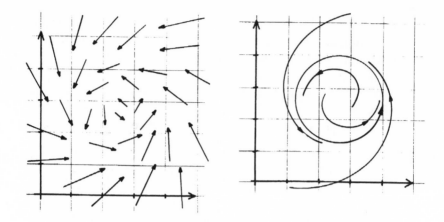

Figure 2.5. A vector field having a set of attracting points, and its trajectories.

Just to fix our ideas about the long-run behaviors exhibited by systems such as those shown in Figures 2.4 and 2.5, let's look at an example of a familiar real-world process that displays both of these sorts of trajectories.

A Grandfather Clock

The magic that makes a grandfather clock work as a timepiece is the regular, cyclical motion of the bob at the end of the clock's pendulum as it swings back and forth. Regarding such a clock as a dynamical system, the set of states is the collection of all the positions in which the bob can possibly be, together with the speed and direction of motion of the bob when it's at those various locations. Consequently, we can specify the state of the clock at any particular moment by giving two numbers: the first represents the position of the bob away from its rest point and the second its velocity, say, positive for motion to the right and negative if the bob is moving to the left. Thus, the state manifold M for this dynamical system is simply the two-dimensional plane R^2, the points of the plane representing the position and velocity of the clock's bob at any moment t (more precisely, a bounded subset of the plane because the bob's position and velocity are both bounded). For the sake of discussion, let's agree to call the state of the clock when the bob is at rest the zero state.

Assume at first that there is no friction in the clock's mechanism (i.e., the pendulum is "undamped"). So when we pull the pendulum arm away from the zero state and let it go, the bob swings to and fro, tracing out the same arc forever. The amplitude of this arc is determined solely by the distance the bob is pulled away from the rest state. The trajectory of the bob in this no-friction case is shown geometrically in Figure 2.6. The picture makes it clear that the long-run behavior of the bob in this undamped case is simply a circle in the space of possible states, which are shown in the right half of the figure. So in the absence of friction, the motion of the bob is periodic.

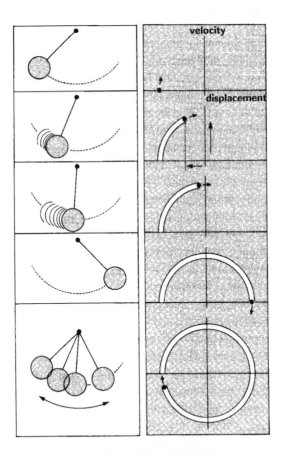

Figure 2.6. Periodic motion of a grandfather clock with no friction.

Now let's forget about idealized frictionless clocks and sprinkle a little dust and sand onto the bearings of the clock's mechanism. In this case, the clock will run down, with the bob eventually returning to the system's rest state at the origin. Thus, the final state for this "damped" system is a single point at which the system's vector field vanishes. Figure 2.7 shows the way the clock's pendulum spirals down to this point.

A set of points transformed back to points in the same set by the action of the dynamical system's vector field is called an *invariant set* of the system. With a couple of additional technical conditions that we won't go into here, an invariant set is also an *attractor* for the system's vector field. Figures 2.6 and 2.7 show that there are at least two different types of attractors: an *equilibrium point* and a *periodic orbit*. The first, of course, is just a single point, or state, at which the vector field vanishes. So once you're at an equilibrium point, the system simply stops. The second type of attractor is a closed curve. This is a set of points that the system visits and revisits forever. These are the classical types of attractors, which were almost certainly known to Newton. They constitute the long-run behavior of many important systems. For example, the radioactive decay process discussed earlier has an equilibrium point

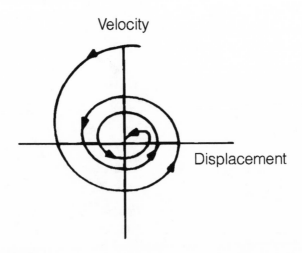

Figure 2.7. Long-run behavior for a real grandfather clock.

at 0. Once the trajectory reaches this state there is no more decay and the process just stops. But it takes an infinite amount of time to actually get to this level!

Over the past few decades, mathematicians have discovered two other types of attractors, leading to all the excitement in dynamical system theory culminating in the "chaos revolution." Before turning to a consideration of these "New Age"-type attractors, let's note in passing that the Poincaré-Bendixson Theorem says that if we have a "well-behaved," continuous-time dynamical system moving in the plane, then the only type of attractors it can have are equilibrium points and periodic orbits. This result effectively rules out nonclassical attractors for all but pathological types of planar dynamical systems (e.g., those with discontinuous vector fields). So to obtain other types of attractors, we have to consider systems whose state manifold is either of dimension greater than two, or has some topological characteristics, such as holes, distinguishing it from a plane. Let's begin with the second possibility.

Consider the two-dimensional torus shown in Figure 2.8, which can be obtained from the square on the left by first folding the square into a cylinder and then gluing the two ends of the cylinder together. Mathematically, these two operations involve first identifying the lines $y = 0$ and $y = 2\pi$, and then making a similar identification of the lines $x = 0$ and $x = 2\pi$. The vector field for this two-dimensional system is given by the rule $v(x, y) = (v_1(x, y), v_2(x, y))$.

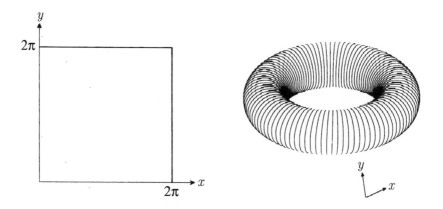

Figure 2.8. A square in the plane and its associated two-dimensional torus.

Now let's take the torus of Figure 2.8 to be the state manifold of a dynamical system, whose vector field is given by the rule $v_1(x, y) = \omega_1$, $v_2(x, y)y = \omega_2$, where ω_1 and ω_2 are fixed real numbers. In particular, suppose we set $\omega_1 = 2$ and $\omega_2 = 3$. The trajectory of this system in the plane, together with its image on the torus, is shown in Figure 2.9. It is clear from this setup that the trajectory is a closed curve. And, in fact, it turns out that this is always the case when the quotient ω_1/ω_2 is a rational number (i.e., the ratio of two integers). This situation should be compared with that shown in Figure 2.10, in which $\omega_1 = 1$ and $\omega_2 = \pi$, so that the quotient is an irrational number. In that case, the trajectory never closes back on itself and, in fact, fills up the entire torus. This is an example of an attractor called a *quasiperiodic orbit*. Such attractors often occur in systems in which there is some quantity, such as energy, that is conserved because in such systems the state cannot wander off to infinity and thus generate infinite amounts of energy. In passing, we should note that the situation described here in two dimensions involving the rational/irrational distinction can be extended to n dimensions at the expense of introducing a more general notion of what it means for n real numbers to be rationally dependent. The interested reader should consult the material cited in the references for more information about this higher-dimensional situation.

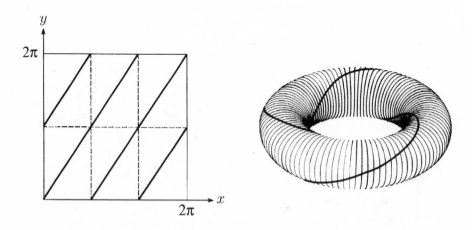

Figure 2.9. Dynamical system on a torus with rational dependence on parameters.

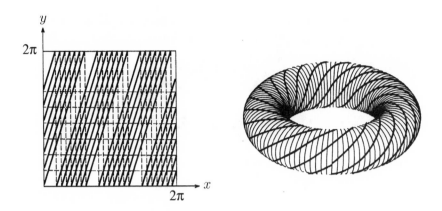

Figure 2.10. Dynamical system on a torus with irrational dependence on parameters.

The final type of attractor is the most complicated and, surprisingly, the most common. It is what is called a *strange attractor*. Roughly speaking, a strange attractor is a lot of periodic orbits and aperiodic trajectories all rolled up into one big tangled mess. A good visual image of such a monstrosity is a ball of yarn or, maybe even better, a bowlful of spaghetti (with sauce). Each strand of spaghetti in the bowl is one part of the strange attractor, the spaghetti sauce ensuring that no two strands ever quite make direct contact. Yet on each strand there is some point that is as close as we like to any other strand, some strands even closing back on themselves to become periodic orbits. But to be on a strange attractor, periodic orbits must generally be of a special type: unstable. This means that the path describing the orbit acts like a repeller rather than an attractor. So if we move off such an orbit by even a little bit, we get pushed even farther away to another part of the attractor instead of being attracted back to the original orbit—as would be the case if the orbit were stable. A strange attractor is shown schematically in Figure 2.11. We defer discussion of the behavior of systems possessing strange attractors to a later section.

For now, let's turn our attention to another crucial aspect of dynamical systems, the stability of both the attractors and the systems themselves. Stability is important because systems are often perturbed away from their attractors. So we want to know when such a perturbation will be damped out rather than grow to a level that destroys the system. Stability theory helps us answer this type of question.

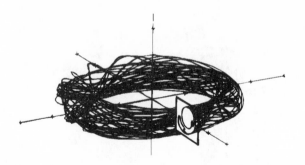

Figure 2.11. A strange attractor.

Stability

Remember the grandfather clock? Suppose we foolishly pull the pendulum bob all the way to the top, so that it stands precariously balanced in a position exactly opposite to that of the zero state. What is the situation with the attractor now? Herein lies an important tale in the storybook of dynamical systems.

When we push the pendulum bob just a little bit away from its equilibrium position at the top of its swing, common sense and experience tell us that it will immediately begin to swing on a long downward arc and ultimately wind up in the zero state (assuming there is some friction)—exactly the same final behavior as that observed when the bob is perturbed slightly from the zero state. So in both cases the bob ends up at the bottom. This kind of behavior leads us to term the state at the bottom as *stable equilibrium* because small (or in this case even large) disturbances away from this state eventually damp out and the bob finally returns to this state. By way of contrast, the equilibrium state at the top is called *unstable* because small departures from it lead the bob to end up not at the top again, but at the bottom. So a stable equilibrium point is one that "sucks back" any trajectory that starts nearby; an unstable one, on the other hand, pushes a nearby state farther away.

As a first step toward mathematically characterizing whether a given equilibrium state is stable or not, let's consider the simplest type of dynamical system, one whose dynamical law of motion is linear. Suppose we have the one-dimensional, discrete-time linear system $x_{t+1} = ax_t$, where t is time, a is a real number, and the system's initial state is x_0. Clearly, the state $x^* = 0$ is an equilibrium for this system (the only equilibrium state, actually) because if $x_t = 0$, then $x_{t+1} = 0$ for all times t. Is it stable, unstable, or neither? Let's see.

It's a simple matter to verify that the trajectory for the above system is given by the expression $x_t = a^t x_0$. Therefore, any small perturbation x_0 away from the equilibrium at the origin will result in the trajectory x_t becoming unbounded as t becomes large if and only if the magnitude of a is larger than 1; but if a is less than 1, the perturbation x_0 is damped out, and the system state eventually returns to the origin. Of course, when $a = 1$, the state remains at x_0 forever, and if $a = -1$, the system oscillates back and forth between the states x_0 and $-x_0$ forever. We can summarize this entire situation by saying that the equilibrium at the origin is stable if and only if $|a| < 1$, unstable if and only if $|a| > 1$, and neither when $|a| = 1$. If we had begun with the continuous time system described by the vector field $v(x) = ax$ instead, the system trajectory would then be given by the rule $x(t) = e^{at} x(0)$. This leads to the conclusion that the equilibrium at the origin is stable for the continuous time system if and only if a is negative.

The foregoing stability results for one-dimensional linear systems can be extended to n-dimensional, continuous time processes in the following way. We consider the system with a vector field given by

$$v(x) = Ax(t), \quad x(0) = x_0, \tag{†}$$

where $x(t)$ is a point in the n-dimensional space R^n and A is an $n \times n$ constant matrix. Again, the only equilibrium for this system is the state $x^* = 0$. The central role in determining the stability of the point at the origin is played by the *eigenvalues* of the matrix A. These are numbers λ satisfying the equation $Ax = \lambda x$ for nonzero vectors x. Generally speaking, these are complex numbers of the form $\lambda = a + bi$, where a and b are real numbers and $i = \sqrt{-1}$. Those λ having $a < 0$ form the left half of the complex plane, whereas the right half is formed by λ having $a > 0$. Numbers with $a = 0$ form the imaginary axis.

We then have the

LINEAR STABILITY THEOREM   *The equilibrium at the origin for the system* (†) *is stable if and only if the eigenvalues of the matrix* A *all lie in the left half of the complex plane.* ∎

The analogous result for the n-dimensional, discrete time system $x_{t+1} = Ax_t$ is given by the following corollary.

COROLLARY *The equilibrium at the origin is stable for the discrete time linear system if and only if the eigenvalues of the matrix* A *all lie inside the unit circle in the complex plane. That is, the set of* λ *whose distance from the origin is less than 1.* ∎

In summary, the linear, continuous-time system is stable if and only if the eigenvalues of A all have negative real parts; the discrete time system is stable if and only if the eigenvalues of A lie at a distance less than 1 from the origin. Now we know when a perturbation in a linear system dies out over time. The answer is completely determined by how the eigenvalues of A are placed in the complex plane. This theorem, along with its corollary, invite two obvious questions:

1. What happens when one of the eigenvalues of the matrix A lies on the imaginary axis?
2. To what degree do the results of the Linear Stability Theorem still hold for nonlinear dynamical systems that are "almost linear"?

The answer to the first question involves a change in the nature of the system attractor from an equilibrium point to a periodic orbit, at least for two-dimensional systems. The second question is answered by the Hartman-Grobman Theorem. Finally, the celebrated Center Manifold Theorem, which we take up in a later section, ties the two questions together. So let's consider these important pieces in the stability mosaic in turn.

Periodic Orbits

The starting point for addressing question 1 is a planar linear system. If the two eigenvalues of system matrix A lie on the imaginary axis, then the coefficient matrix A for such a system has the form

$$A = \begin{pmatrix} 0 & a \\ -a & 0 \end{pmatrix},$$

where a is a real number. More specifically, the eigenvalues of this matrix are the purely imaginary numbers $\lambda = \pm ia$, which by a straightforward calculation leads to the trajectories of the system as being given by the expression

$$e^{At}x(0) = \begin{pmatrix} \cos at & -\sin at \\ \sin at & \cos at \end{pmatrix} x(0).$$

Therefore, if $x(t)$ is a trajectory of the system, so is $x(t + 2\pi/a)$. In other words, the system trajectories are periodic orbits whose period is inversely proportional to the size of the parameter a. These trajectories form a set of concentric closed orbits in the plane, as shown in Figure 2.12.

Each of the closed orbits in Figure 2.12 is an attractor for the system—but one that is neither stable nor unstable, at least not in the sense discussed earlier. Suppose we choose a starting point $x(0)$ that is on one of the closed orbits. A small perturbation of this starting point gives rise to a trajectory that does not eventually come back to the original closed orbit. But it doesn't go flying off to infinity, either. Rather, it just stays on a new periodic orbit (a circle) that is near the original one. How near depends upon the magnitude of the perturbation away from $x(0)$.

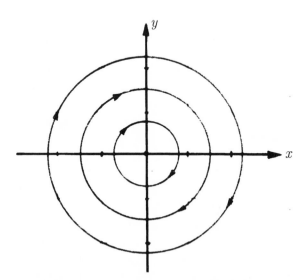

Figure 2.12. Trajectories for a linear system with purely imaginary eigenvalues.

For these reasons, each of the closed orbits shown in Figure 2.12 are called *neutrally stable.* And to distinguish this from the case discussed earlier, in which the perturbation eventually "washes out," we term this earlier type of equilibrium *asymptotically stable.*

The considerations just presented for planar linear systems carry over to higher-dimensional cases as well, again by asking if any of the eigenvalues of the matrix A lie in the right half of the complex plane. If so, the equilibrium at the origin is unstable; if not, it is asymptotically stable. Because it takes only an infinitesimally small perturbation in the entries of the system coefficient matrix A to move any eigenvalues on the imaginary axis into either the left or right half of the complex plane, it is atypical, or *nongeneric,* for a linear dynamical system to have a periodic orbit. This is because periodic orbits arise only out of the eigenvalues of A that are on the imaginary axis. But this situation changes dramatically when it comes to nonlinear processes, for which periodic attractors may be either asymptotically stable or unstable, just as with point equilibria. Now that we know what happens when an eigenvalue of A is on the imaginary axis, let's turn our attention to the second of the two big questions posed earlier: Under what conditions can we extend the stability results for linear systems to systems that are "close" to being linear?

"Almost Linear" Dynamical Systems

To illustrate exactly what we mean when we say a vector field $v(x)$ is "almost linear," suppose we have a dynamical system whose state manifold M is R^n. Let's further assume that the system has an equilibrium point at the origin. Thus, the vector field $v(x)$ describing this system is such that $v(0) = 0$. Moreover, let's assume that the vector field is "smooth," so that it makes sense to write down a Taylor series expansion for $v(x)$ about the origin. This expansion gives $v(x) = Ax + g(x)$, where A is an $n \times n$ real matrix and $g(x)$ is a remainder function with terms of degree 2 or higher. We then say that this system is "almost linear" if the nonlinear terms $g(x)$ are small if x is small. To make the notion of "small" precise takes just a bit more technical virtuosity than we care to go into here. But it can be done. This means that near the equilibrium point at the origin, the behavior of the system is governed by its linear part, that is, $v(x) \approx Ax$ for x small.

The Hartman-Grobman Theorem states (roughly) that if you stay close enough to an equilibrium point, the behavior of a nonlinear system doesn't differ from that of the system's linear approximation—provided that the linear approximation has no eigenvalues on the imaginary axis. In that case, the system state space can be divided into two parts: the stable and unstable manifolds, corresponding to those initial points that ultimately go back to the equilibrium or move away from it, respectively. Now we can phrase our basic question somewhat more precisely: Under what circumstances do the trajectories of the nonlinear system described by the vector field $v(x)$ behave like those of its linear approximation (given by the vector field Ax) near the origin?

To answer this question, we have to be a bit more specific about what we mean by the trajectories of the two systems behaving "like" one another. There are several mathematical ways to express this informal idea, some of which we have already examined in Chapter 3 of *Five Golden Rules* when we spoke about two smooth functions being "like" each other. In the current context, a reasonable way to formalize the idea is to say that the two systems are equivalent if there is a continuous, one-to-one, and invertible mapping h from R^n to R^n by which (1) h takes one set of trajectories onto the other, and (2) h preserves the orientation of the trajectories. Such a mapping is technically termed a *homeomorphism*. The basic idea is shown in Figure 2.13, where trajectories of the vector field $v(x)$ in a neighborhood U of the origin are mapped onto those of the system's linear approximation Ax in a neighborhood $h(U)$.

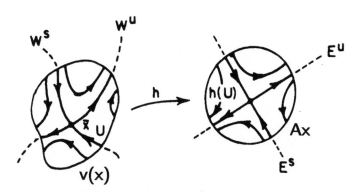

Figure 2.13. A homeomorphism between a system and its linear approximation.

The picture in Figure 2.13 suggests that the two spaces labeled W^s and W^u in the left half of the diagram are mapped onto corresponding spaces E^s and E^u of the linearized version of the dynamical system. To understand what this means, let's describe these spaces.

The space W^s is called the *stable manifold* of the vector field $v(x)$. It consists of all points x near the origin in R^n, such that if the system starts at the point x, the trajectory eventually returns to the equilibrium at the origin. In short, W^s is the set of all small perturbations away from the equilibrium that eventually come back to the origin. Similarly, the *unstable manifold* W^u consists of all those perturbations that do not return to the equilibrium at the origin. The spaces E^s and E^u are, respectively, the stable and unstable manifold of the linear approximation to $v(x)$, which we have denoted Ax. From what has gone before, these spaces are just the eigenspaces associated with the eigenvalues of A that lie in the left and right half of the complex plane, respectively. With these concepts in mind, we can finally state the main result, giving conditions under which a dynamical system is equivalent to its linear approximation near the origin.

HARTMAN-GROBMAN THEOREM *If the linear approximation Ax to the vector field $v(x)$ has no zero or purely imaginary eigenvalues ($\lambda = bi$, b real), then there is a homeomorphism $h(x)$ defined on some neighborhood U of the origin in R^n that takes trajectories of the system $v(x)$ to those of its linear approximation Ax. The map h preserves the orientation of the trajectories, and can be chosen so as to also preserve the time along the trajectories.* ■

Note that preservation of the orientation of the trajectories means merely that a trajectory moving in one direction is always mapped to a trajectory moving in the same direction, whereas preservation of time means that a point on a given trajectory at a particular time is always mapped to a point on the corresponding trajectory at that same instant of time.

Equilibrium points of the type discussed here in which the matrix A has no zero or purely imaginary eigenvalues are called *hyperbolic* equilibrium points. Such points play an important part in what follows because the Hartman-Grobman Theorem says only when the equilibrium is not hyperbolic do the nonlinear terms have to be accounted for in analyzing the stability properties of the equilibrium point.

As an illustration of this fact, consider the planar dynamical system whose vector field is given by

$$v(x) = Ax + g(x),$$

$$= \begin{pmatrix} 0 & 1 \\ -1 & 0 \end{pmatrix} \begin{pmatrix} x_1 \\ x_2 \end{pmatrix} + \epsilon \begin{pmatrix} 0 \\ x_1^2 x_2 \end{pmatrix}.$$

Here, matrix A has eigenvalues $\pm i$. For any nonzero ϵ, the equilibrium at the origin is not a center, as it is with the linear approximation, but a spiral-like attracting point if ϵ is negative and a repelling point if ϵ is positive.

The preceding discussion has moved from a consideration of the stability properties of equilibria—points or periodic orbits—to what happens when we perturb the vector field itself. This latter sort of perturbation has many more ramifications than just telling us when a system is equivalent to its linear approximation. In fact, the perturbation of a vector field is perhaps the key issue about which the theory of dynamical systems revolves, principally because it is the mathematical translation of a crucially important physical question: How does the behavior of a real-world process change when we change the parameters describing the process? Here the process could be the behavior of a national economy governed by parameters such as interest rates or labor availability. Or it may be the dynamical behavior of the human heart, subject to parametric variation in, for example, potassium or sodium concentrations in the blood. In all cases, what we want to know is whether there are critical values of the parameters at which the system's qualitative behavior shifts abruptly from one mode to another. Answering this kind of question is the province of the theory of bifurcations of vector fields. Before taking up these matters, let's quickly pass through an important way station telling us about what the system's behavior looks like near a nonhyperbolic equilibrium point. This is important because it is the first case when we *must* explicitly account for the nonlinear terms in a system.

At the Center of Things

The situation surrounding the Hartman-Grobman Theorem suggests that there is something special about systems having nonhyperbolic equilibrium points, and that for such cases there may well be a third manifold of initial points for which the nonlinear terms cannot be neglected in analyzing the behavior of the nonlinear system—even locally. The possibility of these three types of manifolds, of which the third is called a *center manifold,* is illustrated in Figure 2.14. In this figure, the components labeled E^s, E^u, W^s, W^u are as before, and E^c and W^c consist of those states that emerge from the eigenvalues of the linear approximation to the system that lie on the imaginary axis, and the analogous states of the full nonlinear system, respectively. The Center Manifold Theorem, introduced next, formalizes this splitting of the state manifold.

Suppose we are given the system

$$\dot{x} = Ax + f(x, y),$$
$$\dot{y} = By + g(x, y), \qquad\qquad (**)$$

where x is in R^n, y is in R^m, and A and B are constant matrices of the appropriate sizes. For simplicity, assume further that all of the eigenvalues of B are in the left half plane, whereas those of A all lie on the imaginary axis. In addition, let f and g be smooth functions whose Taylor series begins with terms of second-degree or higher. Our interest is

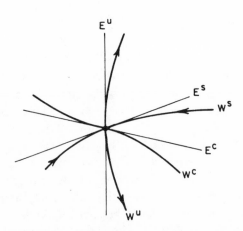

Figure 2.14. Stable, unstable, and center manifolds.

in examining the influence of the x variables on the overall behavior of the system near the equilibrium point at the origin.

Assume now that we have a smooth function h such that $y = h(x)$ is an invariant manifold for the system; that is, points on the surface $y = h(x)$ are taken to new points on this surface by the vector field of the system. We call h a *center manifold* if the Taylor series for h begins with terms of order 2 or higher. Note that if $f = g = 0$, all solutions tend exponentially fast to solutions of $\dot{x} = Ax$. So the behavior of the system on the center manifold determines the asymptotic behavior of the overall system, up to exponentially decaying terms (the y part of the system). The Center Manifold Theorem enables us to extend this argument to the case when f and/or g are not zero.

CENTER MANIFOLD THEOREM *(1) If* x *is sufficiently small, then there exists a center manifold* y = h(x) *for the system* (∗∗). *The behavior of* (∗∗) *on the center manifold is governed by the equation*

$$\dot{u} = Au + f(u, h(u)). \tag{‡}$$

(2) The equilibrium solution of (∗∗) *has exactly the same stability properties as the equilibrium solution of* (‡). *Furthermore, if the equilibrium solution of* (‡) *is stable and if* x(0), y(0) *are sufficiently small, then there exists a solution* u(t) *of* (∗∗) *such that as* t *becomes large,*

$$x(t) \rightarrow u(t),$$
$$y(t) \rightarrow h(u(t)). \quad \blacksquare$$

As an example of this key result, techniques given in the chapter references enable us to show that the system

$$\dot{x} = xy + ax^3 + by^2x,$$
$$\dot{y} = -y + cx^2 + dx^2y,$$

whose linear part has an eigenvalue at 0, has a center manifold $h(x) = cx^2$ + terms of order 4 and higher. Thus, the equation governing the stability of the original system is

$$\dot{u} = uh(u) + au^3 + buh^2(u),$$
$$= (a + c)u^3 + \text{ terms of order 5.}$$

Thus, the origin is a stable equilibrium for the original dynamical system for $a+c < 0$ and an unstable equilibrium for $a+c > 0$. But if $a+c = 0$, we need to take more terms into account in our approximation of the center manifold $h(x)$ to pin down the stability properties of the origin.

In passing, it is worth noting that the Center Manifold Theorem can be extended to the case of discrete time systems, as well as to the situation in which the linearized part of $(**)$ may have eigenvalues in the right half-plane. These matters are taken up in great detail in the material cited in the references.

So, the Center Manifold Theorem shows us that the question of whether initial states lying on the center manifold are attracted back to the equilibrium cannot be answered by appealing solely to the information contained in the linear approximation to the system—even for initial states arbitrarily close to the equilibrium. Such a question is an intrinsically nonlinear affair. Now let's raise the ante.

Forks in the Flow

The term *bifurcation* was introduced early in this century by the famed French mathematician Henri Poincaré to refer to the "splitting" of equilibrium solutions in a family of vector fields. If $v(x, \mu)$ is such a family that depends on the parameter(s) μ, then the equilibrium solutions are given by the solutions of the equation $v(x, \mu) = 0$. As the parameter μ changes smoothly, the values of x solving this equation also vary smoothly for all values of μ—except those for which the linear part (with respect to x) of the vector field $v(x, \mu)$ has an eigenvalue on the imaginary axis. At such distinguished values of μ and x, say at (x^*, μ^*), several branches of equilibrium points may coalesce, in which case we call (x^*, μ^*) a *bifurcation point*.

To illustrate the above idea, consider the vector field $v(x, \mu) = \mu x - x^3$. Here, the linear part is simply the derivative of v with respect to x: $\mu - 3x^2$, which has the single bifurcation point $(x^*, \mu^*) = (0, 0)$. It's a straightforward exercise to verify that the equilibrium point $x = 0$

is stable for all nonpositive values of μ, and that it becomes unstable for positive values. At the bifurcation value $\mu^* = 0$, two new stable equilibria appear, having the values $x^* = \pm\sqrt{\mu}$. This overall situation is shown in the bifurcation diagram of Figure 2.15.

Recall from our discussion of smooth functions in Chapter 3 of *Five Golden Rules* that a one-parameter family of vector fields is a curve in the space $\mathcal{V}$ of all smooth vector fields. Furthermore, such a family typically intersects the unstable classes in a point lying on a surface of dimension one less than the dimension of $\mathcal{V}$. That is, the expected, or "generic," way for such a family to undergo a qualitative change in behavior is by going through what is called a *codimension 1 bifurcation*. There are two standard ways in which this can happen. Let's describe them in the case in which the parameter μ is just a single real number.

The first situation occurs when for a certain value of μ, say $\mu = \mu^*$, the linear part of the vector field has a pair of eigenvalues on the imaginary axis. This situation leads to what is termed the *Hopf bifurcation*. The second case turns up when the linear part of the system has 0 as a simple eigenvalue. This is the famous *saddle-node bifurcation*. In both these cases, the Center Manifold Theorem assures us that locally (in (x, μ)-space) near the bifurcation point at $(0, \mu^*)$, the canonical saddle-node dynamics assumes the form

$$\frac{dx}{dt} = \mu^* - x^2, \qquad \frac{dy}{dt} = A_+ y, \qquad \frac{dz}{dt} = A_- z,$$

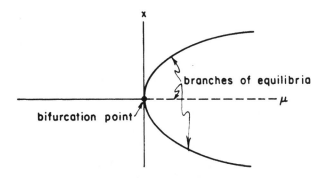

Figure 2.15. Bifurcation diagram for the vector field $v(x, \mu) = \mu x - x^3$.

where $A_\pm$ are matrices having k and ℓ roots in the right and left half-planes, respectively. Thus, the only locally nonlinear behavior comes from the one-dimensional "x-part" of the dynamics.

Similarly, if in the linear part of the system a complex conjugate pair of roots crosses the imaginary axis at $(0, \mu^*)$, then the standard form of the dynamics is

$$\frac{dx}{dt} = -y + x \left(\mu^* - (x^2 + y^2) \right),$$

$$\frac{dy}{dt} = x + y \left(\mu^* - (x^2 + y^2) \right),$$

$$\frac{dw}{dt} = B_+ w,$$

$$\frac{dv}{dt} = B_- v,$$

where again the matrices $B_\pm$ represent the linear parts of the dynamics coming from the subspaces corresponding to the eigenvalues of the linear part of the system lying in the left and right half-planes.

The importance of the saddle-node bifurcation is that all bifurcations of one-parameter families at an equilibrium with a zero eigenvalue can be perturbed to saddle-node bifurcations. So we can expect that bifurcations involving a 0 eigenvalue that we meet in practice will be saddle nodes. If they are not, there is probably something special about the problem that restricts the context to prevent the saddle node from occurring. A typical type of constraint of this sort arises when there is some sort of symmetry in the problem that forces the eigenvalues of the linear part to be symmetric with respect to the imaginary axis. But if the linear part of the system is not constrained by such considerations, then the generic way that an eigenvalue can cross the imaginary axis is either for a real eigenvalue to go through 0 or for a complex conjugate pair of eigenvalues to cross the imaginary axis. The first case leads to the saddle node; the second to the Hopf bifurcation. We can formalize the second part of these results in the celebrated Hopf Bifurcation Theorem.

HOPF BIFURCATION THEOREM   *Suppose the linear part of the system has a single pair of complex conjugate eigenvalues crossing the imaginary axis with nonzero speed as the parameter μ passes through the value μ^*. Then the stable equilibrium at the origin becomes unstable at $\mu = \mu^*$, and a stable limit cycle is born whose radius is proportional to $\sqrt{\mu - \mu^*}$.* ∎

As far as the qualitative structure of the trajectories goes, in the case of the Hopf bifurcation we have a stable equilibrium point becoming unstable, together with the birth of a stable limit cycle. For the saddle node, we have either the appearance or disappearance of a stable and unstable equilibrium, depending upon the direction in which the parameter μ passes through the critical value μ^*. The bifurcation diagrams for these two cases are displayed in Figure 2.16.

- *Example: The Ball Bearing and the Basketball:* Consider a small, solid sphere—a ball bearing, for instance—inside a hollow sphere such as a basketball. Now hang the basketball from the ceiling by a cord, and start rotating it at a rate of rotation ω.

If ω is small, the bottom of the basketball is a stable point and the ball bearing is attracted back to it if perturbed away. But when the rate of rotation exceeds a critical value ω_0, the ball bearing moves up the side of the basketball to a new fixed point. (*Note:* The actual value of ω_0 at which this happens depends on the friction between the surface of the ball bearing and that of the basketball.) For each rotation rate ω greater than ω_0, there is a stable, invariant circle of fixed points, as shown in

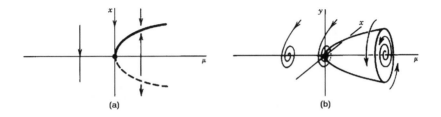

Figure 2.16.   (a) The saddle-node bifurcation, and (b) the Hopf bifurcation.

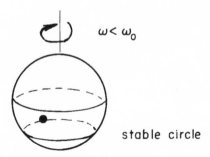

$\omega < \omega_0$

stable circle

Figure 2.17. The ball bearing and the basketball.

Figure 2.17. We get this entire surface of fixed points rather than a single point because of the symmetries of the problem.

From the foregoing analysis, we have seen that there may be an entire surface of equilibria for systems containing parameters. So it may well be of interest to ask about the mathematical nature of this surface. In particular, is the surface smooth? Does it have holes? How does the structure of the surface depend on the mathematical character of the vector field? And so on and so forth. The mathematical theory of catastrophes introduced in Chapter 3 of *Five Golden Rules* is one way of answering some of these questions.

Gradient Dynamical Systems and Elementary Catastrophes

Suppose we have a dynamical system described by a vector field $v(x)$ that can be written as the gradient of a scalar function $V(x)$, that is, $v(x) = \text{grad } V(x)$. For obvious reasons, such systems are called *gradient dynamical systems*. To make contact with the material of Chapter 3 of *Five Golden Rules,* let's further assume that the function $V(x)$ has as many derivatives as we wish. Gradient dynamical systems and catastrophe theory come together when we recognize that the equilibrium points of the dynamical system (the points where $v(x) = 0$ coincide with the critical points of the smooth function $V(x)$, which are the points at which the gradient of $V(x)$ vanishes. Although it is not immediately clear, we should point out that the only type of attractors that gradient systems can have are point equilibria. Thus, it makes sense to associate such attractors with the critical points of the smooth function V.

In view of this coincidence of point equilibria and critical points, it's a pretty good bet that the catastrophe points for the smooth function $V(x)$ have some bearing on the nature of the bifurcation points of the dynamical system described by the vector field $v(x)$, which indeed turns out to be the case. So let's look at this connection in a bit more detail.

Suppose our dynamical system contains parameters that we represent by μ. Then we know that a change in the number of solutions of $v(x, \mu) = 0$ occurs at exactly those values of μ at which the function grad $V(x, \mu) = 0$. Furthermore, from Chapter 3 of *Five Golden Rules* we know that this latter situation is connected with the appearance of catastrophes. Thus, we can consider the general bifurcation problem for such systems as the problem of determining the set of solutions of the equation

$$\text{grad } V(x, \mu) = 0,$$

for some curve

$$\mu(s) = (\mu_1(s), \mu_2(s), \ldots, \mu_k(s)),$$

in R^k, where k is the number of parameters in the system. Even though there are an infinite number of such curves, the behavior of the function $V(x, \mu)$ at any point μ is known from the theory of elementary catastrophes. For instance, if the function V has a cusp type of critical point, then in the neighborhood of this point we know that V has the following canonical form

$$V = \frac{x^4}{4} + \frac{ax^2}{2} + bx$$

for real parameters a and b.

This analysis tells us that the general bifurcation problem for gradient systems reduces to the problem of determining the solution set of the equation grad $V = 0$ for any curve $(a(s), b(s))$ in the space of control parameters, where s is a parameter representing the position along such a curve (technically, the arc length). The solution set is just the collection of points at which the cusp manifold grad $V = 0$ intersects the two-dimensional surface $(x, a(s), b(s))$, whose points are determined by fixing x and s. One such curve and its associated solution set are shown in Figure 2.18. In the figure, the quantity x_c represents the equilibrium point of the dynamical system for the corresponding value of the parameters. We see from Figure 2.18 that there are many values of the parameters yielding more than a single such equilibrium. It is exactly at those points where the solution set becomes multivalued that the critical points of the scalar function V undergo "catastrophic" changes.

The foregoing results point out the importance of gradient dynamical systems if we want to make a tight fit between bifurcation analysis and catastrophes. So we close out this section by giving a simple criterion by which we can decide if any particular dynamical system fits into this especially nice category.

Suppose we are given the parameterized set of differential equations

$$\frac{dx}{dt} = f(x, \mu),$$

and want to know if it is a gradient system. If we let ith component of the vector function f be denoted by $f_i(x, \mu)$, then the criterion for this to be a gradient system is very straightforward and easy to test. A dynamical system is a gradient system for a value $\mu = \mu^*$ if and only if

$$\frac{\partial f_j(x, \mu^*)}{\partial x_i} = \frac{\partial f_i(x, \mu^*)}{\partial x_j},$$

for all i and j. (*Note:* For readers familiar with the terminology of vector analysis, the above relation says simply that the curl of the vector f is zero.)

To illustrate this result, consider the planar system

$$\frac{dx}{dt} = 2xy,$$
$$\frac{dy}{dt} = \mu x^2 - y^2,$$

which arises from the parameterized vector field $v_\mu(x, y) = (2xy, \mu x^2 - y^2)$. By calculating the relevant partial derivatives, we find that this is a gradient dynamical system if and only if

$$\frac{\partial}{\partial y}(2xy) = \frac{\partial}{\partial x}(\mu x^2 - y^2).$$

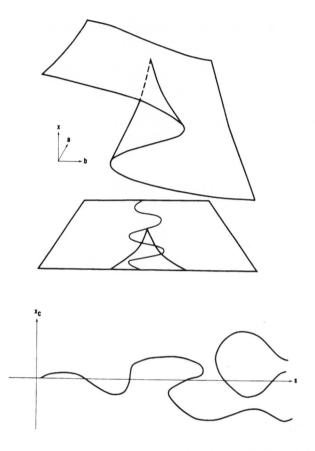

Figure 2.18. A curve in the cusp control plane and its associated solution set.

But this can happen if and only if $\mu = 1$. In that event, the corresponding scalar function is $V(x, y) = x^2 y - \frac{1}{3}y^3$, which incidentally, leads to what we called the elliptic umbilic catastrophe.

The Hopf Bifurcation Theorem, along with the other bifurcation results we have discussed so far, tells us how the classical attractors—equilibrium points and periodic orbits—metamorphose into one another. It gives us detailed information about how and when a stable equilibrium point gives way to a periodic orbit as parameters in the system vary. But earlier we saw that most of the current excitement in dynamical systems revolves about the so-called strange attractors. These are the bizarre objects that give rise to chaotic dynamical behavior. So without further ado, we give the remainder of the chapter over to a discussion of these matters.

A Strangeness in the Attraction

Think of a baker at work rolling out a ball of pastry dough. The baker takes some dough, works it into a thin sheet with a rolling pin, folds the dough over itself a time or two, and then rolls it out again. To make this process mathematically precise, let's assume that at each stage the thickness of the sheet of dough is halved, its length is doubled, and the width remains the same. Instead of folding the sheet, let's assume that the sheet is cut in half, with one of the half-sheets (always the same one) placed atop the other. After this cutting and placing, the process then begins again with a sheet the same size as before. The first two stages of this process are shown in Figure 2.19, where the face of a cat is placed on the sheet to make it easier to envision how the rolling, cutting, and placing are done.

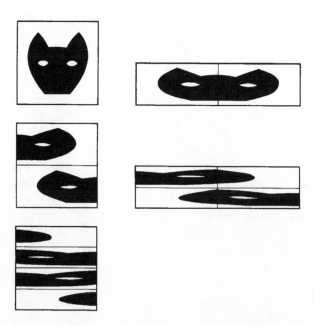

Figure 2.19. Rolling out a sheet of pastry dough.

70

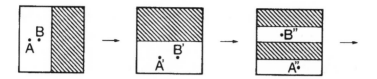

Figure 2.20. The baker's transformation.

Now, instead of a cat's face, let's place two dots, *A* and *B*, on the pastry sheet. Figure 2.20 shows two applications of the baker's transformation in this case. Note in the figure how the points *A* and *B* have separated during the course of successive applications of the transformation. This is a geometric view of what is called *sensitivity to initial conditions:* Two points originally close together get pushed apart by the action of the *baker's transformation.*

The key to what has become known as a chaotic system lies in the two competing operations exemplified by the baker's transformation. These are the processes of stretching the dough by rolling it out, and folding it back over again so that the stretching doesn't expand the dough indefinitely. These two operations are shown geometrically in Figure 2.21; in mathematical terms, the stretching corresponds to a locally unstable equilibrium, one from which all nearby points are repelled, because when we stretch the dough each point is moved away from all its neighbors. The folding operation is the mathematical counterpart of a system whose state space is bounded so that the space can't be stretched everywhere. The reader should keep these two competing processes in mind as we go through the examples of chaos given in the remainder of the chapter.

Figure 2.21. Stretching and folding.

To understand some of the peculiarities of chaotic dynamical processes, let's look at what must surely be about the simplest dynamical system imaginable, a process I call the *Circle-10 system*. This system consists of a particle moving on a circle of circumference 1. The state manifold M of this system is simply this circle, so we can describe the state of the particle at any moment with a number between 0 and 1. Suppose the vector field determining the particle's movement on the circle is given by the rule: "If you're currently at the point x, move to the point $10x$." With this rule, the circle gets stretched to 10 times its length, and is then wrapped 10 times around itself. To keep things as simple as possible, let's assume that the time set for the system is the nonnegative integers, so that the state changes at discrete moments. It turns out that this trivial-looking dynamical system has some very nontrivial properties.

First of all, divide the circumference of the circle into 10 sectors and label them $0, 1, \ldots, 9$. In terms of numbers between 0 and 1, sector 0 corresponds to all numbers between 0 and $0.099999\ldots$; sector 1 runs from 0.1 to $0.19999\ldots$; and so on to sector 9, which consists of the numbers from 0.9 to $0.9999\ldots$. To get things started, suppose the particle begins its journey at the point $x_0 = 0.379762341$. This point lives in sector 3, nearly 80 percent of the way to sector 4. Let's now follow the path of the particle as the Circle-10 rule (its vector field) is iterated (i.e., applied over and over again).

When we apply the state transition rule to x_0 and wrap the circle 10 times around itself, the circle's length expands by a factor of 10. Thus, the point x_0 moves to the point 3.79762341. But note that one time around the circle just gets you back to where you began, and so do two tours, and so do three. So the result of applying the rule is just the same as coming to the point $x_1 = 0.79762341$. This is a point in sector 7. Therefore, on the first iteration the starting point x_0 moves from sector 3 to sector 7. As we continue to apply the vector field in this manner, we generate the results given in Table 2.2.

This listing of the trajectory of the starting point x_0 under the state transition rule of the Circle-10 system shows that the action of this rule is the ultimate in simplicity: Just multiply by 10 at each stage and delete the digit to the left of the decimal point. We can express the effect of this rule even more compactly by "integrating" the dynamics, which yields the closed form for the state of the system at time n as being

Table 2.2. State transitions for the Circle-10 system.

Time	Number from Rule	Point on Circle	Sector
0	0.379762341	$x_0 = 0.379762341$	3
1	3.79762341	$x_1 = 0.79762341$	7
2	7.9762341	$x_2 = 0.9762341$	9
3	9.762341	$x_3 = 0.762341$	7
4	7.62341	$x_4 = 0.62341$	6
5	6.2341	$x_5 = 0.2341$	2
6	2.341	$x_6 = 0.341$	3
7	3.41	$x_7 = 0.41$	4
8	4.1	$x_8 = 0.1$	1
9	1	$x_9 = 0$	0

$x_n = 10^n x_0$ mod 1. (*Note:* The "mod 1" operation is just mathematical shorthand for the process of dropping the digit to the left of the decimal point.)

Examining the sequence of sectors that the particle visits, we find the itinerary from the starting point x_0 consists of the sectors 3, 7, 9, 7, 6, 2, 3, 4, 1, 0, 0, ... If these numbers look familiar, they ought to—they're just the decimal digits of the starting point x_0 itself! This is no accident. For *any* initial point x_0, the itinerary of sectors visited will match exactly the decimal digits of that point for the simple reason that the rule, "Multiply by 10 and chop off the digit to the left of the decimal point," corresponds to nothing more than shifting the decimal point one position to the right. It's difficult to imagine a more straightforward, easy-to-calculate, deterministic dynamical system than this one.

But simple to describe and calculate doesn't necessarily mean simple in behavior. It turns out that the Circle-10 system captures just about all the interesting behavioral features of the sort of systems that have given rise to the theory of chaotic processes. Let's look at these characterizing "fingerprints" of chaos in a bit more detail.

- *Divergence:* Just for fun, consider Champernowne's number

$$C = 12345678910111213141516\ldots,$$

which is obtained by writing down the positive integers in order. Now suppose the decimal digits of the starting point in our Circle-10 system agree with Champernowne's number in the first 100 million places, but thereafter continue with ... 33333 ... forever. Call this initial state c. Thus, the itinerary of the system starting from Champernowne's number C agrees with the trajectory starting from c for the first 100 million steps. But thereafter the itinerary from c stays put in sector 3 forever, whereas that from C goes its merry way, whatever that might be, but it certainly is not complete confinement to sector 3.

This example shows that two starting points C and c, that are closer together than we could ever hope to measure, end up following completely independent paths. More formally, we say that the itineraries *diverge,* and that dynamical systems with this kind of divergent behavior are displaying the kind of sensitivity to initial conditions shown earlier in the baker's transformation.

- *Randomness:* Suppose we record the numbers as they come up on a roulette wheel at the local casino (with 37 slots labeled 0, 1, 2, ... , 36, and 00), and we obtain a sequence such as 5, 23, 12, 30, 2, 18, 4, 17, 31, 24, 1, 00, 11, 32, 25, 17, 27, and 33. Intuition suggests that this sequence is about as random as a sequence can be, at least if it is a fair roulette wheel. Just as we formed Champernowne's number by writing down the positive integers in order, we can form a random number by writing down this roulette sequence in the same way, putting a decimal point in front of it to get a number between 0 and 1. For this sequence of spins of the wheel, we thus get the random number $r = 0.5231230218417312410011322517273 3$. In Circle-10 terms, this is a point on the circle in sector 5. So if we start the Circle-10 dynamical system from this point r, the system will visit exactly the same sequence of sectors as the numbers that turned up on the roulette wheel. But remember that this is a random sequence of numbers. Yet we have succeeded in generating it from what to all appearances is a totally deterministic dynamical system. What this experiment shows is that a deterministic rule of state transition can lead to a sequence of states that's every bit as random as the outcome of the spin of a roulette wheel.

Almost *every* real number has a decimal expansion that's random. Therefore, the purely *deterministic* Circle-10 dynamical system behaves in a random manner not just for a few special initial states, but for almost every starting point.

- *Instability of Itineraries:* If almost all itineraries are random, it is of considerable interest to ask which ones are not. That is, which starting points are periodic, leading to itineraries that repeat themselves over and over again? The answer is clearly those starting points whose decimal expansion is either finite, like our test sequence, or repeats with a finite period (technically speaking, they're actually the same thing). It is a well-known mathematical fact that a number will have such a repeating expansion if and only if it is rational, that is, if it is a number such as 1/2, 7/16, or 207/399, each of which is the ratio of two positive integers.

The interval between 0 and 1 contains an infinite set of rational numbers, as well as a much larger infinity of irrational numbers such as $\sqrt{3}$, π, and Champernowne's number. It can be shown that between any two irrational numbers there is a rational one, although they do not alternate (almost all numbers are irrational). Thus, the starting points leading to periodic orbits are totally mixed up with those *aperiodic points* that do not lead to such cyclic trajectories. This fact also shows that the periodic points are unstable, because if we move just a little bit to a nearby irrational starting point, we're led to an aperiodic trajectory. It turns out to be the case that all but possibly one of the system trajectories are unstable. So regardless of whether we start at a periodic or an aperiodic point, the itinerary is almost sure to be unstable.

The set of itineraries followed by the starting points on our Circle-10 system is an example of a strange attractor. Such an attractor is characterized by the properties just discussed: instability of almost all trajectories, deterministic randomness, and sensitivity to initial conditions. Furthermore, it turns out that most dynamical systems have a strange attractor for some region of the parameter values describing the system. In the more general setting of a continuous-time dynamical system, a typical strange attractor looks something like the object shown earlier in Figure 2.11.

The problem with strange attractors and, hence, chaotic systems is that they may be difficult to distinguish from systems having unstable periodic orbits of very long periods. Of course, from a practical point of view the distinction may not matter—but, then, sometimes it does. In any case, we would certainly like to have a sure-fire test that separates the imposters from the real thing. Fortunately, there are a number of tests that do exactly that. These tests fall into two distinct categories: those for systems in which the vector field v is given, and those for which all we have available are observations of the system's trajectory at different moments in time. Let's begin by looking at tests of the first type.

Up, Up, Down and Away

Suppose we have a system consisting of a set of entities—insects, people, molecules—whose population level at generation t is given by x_t. Furthermore, let the rule specifying how the population grows as a function of its current level be specified by the vector field f defined on some closed interval of the real line, where f has just a single maximum in this interval (i.e., f is a function with one "hump"). Thus, the population at the next generation is $x_{t+1} = f(x_t)$. Knowing rule f, we can compute the entire population history as soon as we are given the initial population level x_0.

In 1973, mathematicians Tien-Yien Li of Michigan State University and James Yorke of the University of Maryland discovered the following remarkable fact about this kind of system: If there is any sequence of three successive generations in which the population increases for two periods and then returns to (or goes below) its original level, the system is chaotic. This situation is shown graphically in Figure 2.22. In more colorful language, the Li-Yorke result is usually expressed by the title of their famous paper, "Period Three Implies Chaos." This result gives us a very simple test for chaos in scalar systems (those whose states can be represented by a single real number). Unfortunately, the theorem fails for higher-dimensional systems, which call for more delicate tests. Before stating the Li-Yorke result more formally, let's first talk just a bit more about the behavior of systems governed by scalar maps such as f.

In the study of periodic orbits of scalar maps, it is of considerable interest to determine when an orbit of a given period implies the existence of orbits of other periods. A powerful result in this direction was given in the mid-1960s by the Russian mathematician A. Sarkovskii and later greatly extended by others. This result provides an actual test for chaotic behavior.

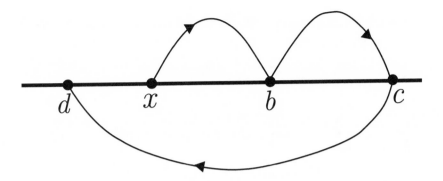

Figure 2.22. The Li-Yorke condition for chaos.

SARKOVSKII'S THEOREM *Construct an ordering of the positive integers according to the following scheme:*

$$3 \to 5 \to 7 \to \cdots \to 3 \cdot 2 \to 5 \cdot 2 \to 7 \cdot 2 \to 9 \cdot 2 \to \cdots \to 3 \cdot 2^2$$
$$\to 5 \cdot 2^2 \to 7 \cdot 2^2 \to 9 \cdot 2^2 \to \cdots \to 3 \cdot 2^n \to 5 \cdot 2^n \to 7 \cdot 2^n \to$$
$$9 \cdot 2^n \to \cdots \to 2^m \to \cdots \to 32 \to 16 \to 8 \to 4 \to 2 \to 1,$$

where the symbol "$\to$" means "precedes." If f is a scalar map having a single global maximum (or minimum) with an initial point x_0 leading to a cycle of period p, then for every q that follows p (that is, for every q such that p $\to$ q), there is an initial point x_0^* leading to a cycle of period q. ■

Note that Sarkovskii's Theorem says nothing about the stability of the various periodic orbits and is only a statement about the cycles emanating from various initial values of x.

 An extremely important corollary of Sarkovskii's result is that if a map has a cycle of period 3, then it has cycles of *all* integer periods. This observation forms the basis for one of the most celebrated results in chaos literature, the famous Period-3 Theorem.

PERIOD-3 THEOREM *Let f be a continuous, "one-humped" map of an interval I to itself. Let x be a point in I such that the first three iterates* *of* x *are given by* $f(x) = b$, $f^{(2)}(x) = c$ *and* $f^{(3)}(x) = d$. *Assume that these iterates are ordered as in Figure 2.22, that is,*

$$d \leq x < b < c \text{ (or } d \geq x > b > c).$$

Then for every positive integer k, *there is an initial point in* I *leading to a cycle of period* k. *Furthermore, there is an uncountable subset* U *of* I *containing no periodic points. In addition, the points in* U *satisfy the following conditions:*

1. *For every pair of distinct points* x *and* y *in* U, *the trajectories starting from* x *and* y *do not intersect, yet come arbitrarily close to each other.*
2. *For every point* x *in* U *and periodic point* x_0 *in* I, *the trajectories from* x *and* x_0 *do not intersect.* ∎

So the Period-3 Theorem improves upon Sarkovskii's result by characterizing some aspects of the aperiodic points. Namely, by (1), every aperiodic trajectory comes as close as we want to another aperiodic trajectory—but they never intersect. Furthermore, from (2), the aperiodic orbits have no point of contact with any of the periodic orbits either. We also note that the conditions of the theorem are satisfied by any function f that has a point of period 3 (although it's not necessary for f to have such a point for the type of chaos described to arise).

It is of some interest to note that the Poincaré-Bendixson Theorem given earlier shows that chaotic behavior cannot occur for continuous-time dynamical systems if the dimension of the system's state space is less than 3. Yet, the Li-Yorke result shows that when we switch to a discrete time system, strange attractors and chaos can appear even for scalar systems. The most famous such example is the so-called *logistic map*, for which the relevant vector field is $f(x) = \mu x(1 - x)$, where μ is a parameter between 0 and 4 (this is the only range of values that keeps the state x bounded between 0 and 1). For this system, a strange attractor emerges for most values of $\mu > 3.57 \ldots$ We'll elaborate on what we mean by "most" in a moment.

The Period-3 Theorem provides a sufficient condition for a mapping f to display chaotic behavior. There are two main features of the strange attractor characteristic of such a system. It has: (1) a countable number of periodic orbits, and (2) an uncountable number of aperiodic orbits. We can add to this the requirement that the limiting behavior is an unstable function of the initial state, that is, that the system is highly sensitive to small changes in the initial state x_0. Now let's turn our attention to the description of another test that can be used to identify systems having such chaotic behavior.

The Lyapunov Exponents

Recall the baker's transformation shown in Figure 2.20. In that pastry-flattening situation, there are two directions in which points in the dough can move: up–down and left–right. So there are two rates of separation, one in the vertical direction and one in the horizontal. These rates are characterized by two numbers, called the *Lyapunov exponents* for the process. If either of these numbers is positive, it means there is a positive rate of separation of the initial points in at least one of the two directions. In that case, we say the system displays sensitivity to initial conditions, or, more commonly, that the system is chaotic. The two-dimensional pastry-rolling setting can be easily extended to the case of an n-dimensional system, in which case there are n such numbers measuring the separation rate of nearby initial points in each of the n directions. Because there is one such Lyapunov exponent for each degree of freedom of the system, we have the kind of sensitivity to initial conditions indicative of chaos if even one of these numbers is positive.

As an illustration of this result, consider the scalar logistic map just discussed. Because $n = 1$ for this system, there is just a single Lyapunov exponent, let's call it λ. Figure 2.23 shows the value of this quantity as a function of the parameter μ characterizing the system's vector field. The values of μ for which the Lyapunov exponent λ becomes negative correspond to regions of periodic—rather than chaotic—behavior. So we see from Figure 2.23 that for values of μ greater than about 3.57, most parameter values lead to chaotic behavior for most initial states. But there are always small "pockets" of values of μ for which some initial states lie on periodic orbits instead of on aperiodic ones. For more details about this map, the reader should consult the chapter references.

An extremely interesting application of the Lyapunov exponent was made in 1989 by the astronomer J. Laskar, who in an extensive

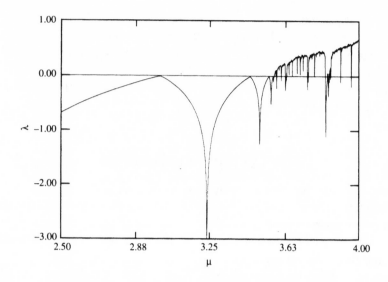

Figure 2.23. The Lyapunov exponent for the logistic map.

series of numerical experiments discovered strong evidence suggesting that the solar system itself is chaotic. Let's see how.

Chaos among the Planets

Following the pioneering work on celestial mechanics by Laplace and Lagrange in the eighteenth century, in which they laid down the basic mathematical rules of planetary motion, people assumed that the attractors of the planetary motions were quasiperiodic orbits (in today's terminology). This led to a considerable amount of work in the nineteenth century trying to construct quasiperiodic solutions for all the planets in the solar system, using methods based on perturbations away from nominal circular orbits. Poincaré shed considerable doubt on these efforts, however, when he showed that if it's possible to express the solutions of the planetary motions as infinite series, then in general these series do not converge.

In the 1950s, however, A. Kolmogorov and V. Arnol'd of Moscow University and J. Moser of New York University proved that if the perturbations are not too large, that is, if the planetary masses and the eccentricities and inclinations of their orbits are "small," then there is a large set of initial conditions that lead to quasiperiodic orbits. But no one has ever shown that the solar system satisfies these conditions.

To settle the matter of whether the initial conditions of the solar system lead to quasiperiodic or chaotic motion, J. Laskar carried out a detailed numerical integration of the trajectories of the outer planets (Jupiter to Pluto) over periods ranging from 500,000 to 1 billion years. To test for chaotic behavior, Laskar calculated the largest Lyapunov exponent for the planetary system. A typical result is shown in Figure 2.24, which plots the quantity $\log_{10}(\log_{10}(d/d_0)/T)$ versus $\log_{10} T$. Here d is the distance between the orbits, $d_0 = 10^{-7}$ is the relative initial separation between the orbits, and T is the time (100 million years in Figure 2.24). Curve 1 shows the Lyapunov exponent when the calculation is renormalized every 40 million years, whereas curve 2 was generated by renormalizing the computation every 1 million years. (Note that the curves do not coincide for the first million years because of different initial conditions.) What is striking about this result is that we obtain essentially the same estimate for the largest Lyapunov exponent for large values of T: about $10^{-6.7} \text{yr}^{-1}$, a quantity that in this context is significantly larger than zero.

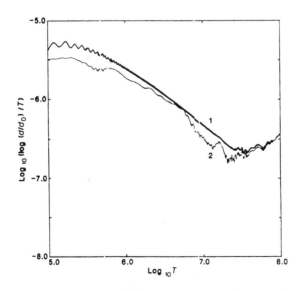

Figure 2.24. Computation of the largest Lyapunov exponent for the solar system.

One outcome of Laskar's calculations is to give us a measure of the limit of predictability of the solar system. This estimate turns out to be

$$d(T) \approx d(0) \exp(T/5),$$

where $d(T)$ is the distance separating two nearby orbits in state space at time T (measured in millions of years). This expression shows that it is possible to predict the planetary motions quite accurately over a period of 10 million years—but not over 100 million years! For instance, an error of as small as 10^{-10} in the initial conditions leads to a 100% error in the estimate of the trajectory after 100 million years. Given that Laskar's mathematical representation of the solar system is a close approximation to the real thing, we can conclude only that the motion of the planets is very likely to be chaotic. Whether this implies catastrophic events such as the Earth's orbit crossing that of Venus remains to be seen.

It's clearly asking a lot to demand that the entire rule of motion for the system be given. In fact, for the vast majority of physical and human systems, all we have at our disposal is a set of observations (that is, measurements) about how the system has behaved in the past. In such cases, the very goal to which theoretical science aspires is to produce an explicit characterization of the rule generating what has been seen (i.e., to find a vector field that could plausibly have given rise to the observations). Obviously, it would be of great interest to know if the kind of rule we're looking for should have a strange attractor or not. This situation leads to the so-called *external,* or outsider's, view of the system, in which we seek tests for chaos based solely upon the measured data. In a chaotic process, the trajectory from almost every starting point wanders over a large fraction of the state space. But not all points of the state space are created equal; some points are visited more frequently than others. This fact provides the basis for the following chaos test based solely on a system's observed behavior.

The Correlation Dimension

Consider for a moment the two classical attractors, an equilibrium point and a limit cycle. Regardless of the dimension of the overall state space, the equilibrium point has dimension 0, whereas the limit cycle, being a simple closed curve, has dimension 1. These numbers are the *geometric dimension* of these attractors. At the other end of the scale is the situation in which the system is truly random. In this case, every point of the state space is eventually visited, leading to the attractor having the same geometric dimension as the state space itself. Chaotic systems, with their strange attractors, lie somewhere in between. For such systems, the attractor is clearly not such a primitive geometrical object as a point or a simple curve. Yet it's still smaller somehow than the state space itself.

To formalize this idea, suppose we look at a trajectory that has been moving on its attractor for some time. Let's sample a set of points from this trajectory, and compute the statistical correlation among these sample values. This calculation results in a quantity called the *correlation dimension* of the system, a number measuring the degree to which the system's attractor "fills up" the space of states. Note carefully that this quantity does not usually equal the geometric dimension of the attractor. This is because the correlation dimension weights the points on the attractor according to how frequently they are visited. The geometric dimension, on the other hand, attaches the same weight to each point of the attractor regardless of how often it is visited. A tell-tale sign of chaos is when the correlation dimension turns out to have a noninteger value substantially greater than 1.

Let's illustrate the general idea by taking data from the logistic map and plotting it in the following way. First take the values at times $t = 0$ and $t = 1$. Call these values x_0 and x_1, respectively, and think of them as the coordinates of a single point in the two-dimensional x-y plane. So the x coordinate of this point is the number x_0, and the y coordinate is x_1. Now do the same thing for the pair of values (x_1, x_2), which are the observed outputs from the logistic rule at times $t = 1$ and $t = 2$, respectively. Continuing in this fashion, we obtain a scattering of points in the x-y plane. Figure 2.25 shows the possibilities. If the original data points came from a truly random process, then we would expect to see the picture shown in part (a) of the figure. In this case the points are scattered more or less uniformly throughout the plane. But our data did not come from a purely random process. Rather, it was generated by a very simple, very deterministic rule. So what we actually see is the picture in part (b). This distribution of points is not random, as it certainly does not fill up the two-dimensional plane. This fact suggests the presence of a strange attractor. The correlation dimension sketched

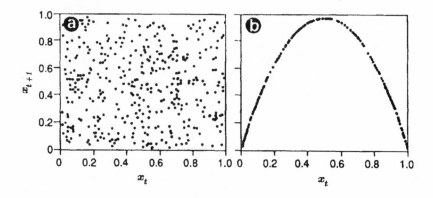

Figure 2.25. (a) A random scatter. (b) The logistic parabola.

earlier is the generalization of this simple idea to settings in which the system's observed behavior lives in an n-dimensional world rather than in the one-dimensional universe of the logistic map.

So with this final test for chaos, let's leave the theory behind for a while and turn our attention to a real-life situation in which these rather abstract notions can be put to productive use.

Gold Bugs and Efficient Markets

In a 1988 study, statisticians Murray Frank and Thanasis Stengos decided to use the fluctuation of the price of gold as a basis for testing one of the most cherished assumptions of financial theorists the world over, the so-called *Efficient Markets Hypothesis (EMH)*.

The EMH comes in many flavors, but the one that interests us here is the weakest version, which sometimes masquerades under the label of the *Random Walk Hypothesis*. In rough terms, the hypothesis asserts that a time series formed of price *differences* is a completely random sequence. In short, knowledge of the past price history of a stock or commodity gives no information of any kind about what the price of that item will be in future time periods. If the random walk theory is true, then, of course, all schemes for beating the market based upon statistical processing of price histories are doomed to failure, at least in the long run; there can be no such scheme for consistently beating the market. Frank and Stengos devised an experiment to test this claim in the gold market by looking for hidden chaotic structure in actual gold price data.

To describe the Frank and Stengos results, let p_t be the price of gold at period t, and define the relative price difference to be $r_t = (p_t - p_{t-1})/p_{t-1}$. The study revolves about the linear regression equation

$$r_t = \beta_0 + \beta_1 r_{t-1} + u_t,$$

for predicting the sequence $\{r_t\}$, where the sequence $\{u_t\}$ is a zero mean, unit variance normal random sequence representing the error arising from the use of this linear relation to estimate the price difference. Statistical theory asserts that if the gold market is indeed efficient, then we should have $\beta_0 = \beta_1 = 0$, which means that the price differences form a normally distributed random sequence.

Using standard statistical techniques on the actual gold price data, the estimates of these coefficients turned out to be $\beta_0^* = 0.0004$ and $\beta_1^* = -0.606$ with a high confidence level. So the standard statistical tests strongly suggest that the gold market is indeed efficient. But not so fast!

The situation changes dramatically when the tests for chaos given above are applied to the same gold price data. By computing the correlation dimension of the data and comparing it with the dimension of a randomly generated series of numbers, we discover that the correlation dimension is not only not an integer; it is consistently on the order of one-half as large as the dimension of the random data. Furthermore, Frank and Stengos estimated something called the K-entropy, which is the sum of the positive Lyapunov exponents. But in this case, the K-entropy and the largest Lyapunov exponent coincide because the dynamical system for the sequence of price differences $\{r_t\}$ is just a scalar system. Thus, if the K-entropy is positive, so is the largest Lyapunov exponent (which here is the only Lyapunov exponent). The estimate that Frank and Stengos came up with for the K-entropy was $K = 0.15 \pm 0.07$, a number significantly greater than zero.

Both of these calculations strongly suggest that there is some underlying nonlinear structure in the gold price data that is not being picked up by the linear methods used to estimate the parameters β_0 and β_1, and that it is highly probable that there is a strange attractor lurking at the heart of the gold market price-setting mechanism (whatever *that* may be).

In the popular literature on chaos, the wonders of chaotic dynamical systems are almost always linked with those most fascinating of geometrical objects, *fractals*. So just about now the reader might well be wondering, What's the connection? How do the strange attractors of dynamical systems make contact with fractal objects such as the Sierpinski gasket or the famed Mandelbrot set? Let's conclude our excursion into the world of systems and dynamics by answering this question.

Domains of Attraction

We spoke earlier about the stable and unstable manifolds of an equilibrium point of a vector field. These were the points in the state space that were, respectively, attracted to or repelled from the equilibrium point in question. For instance, consider the dynamical process described by the differential equations

$$\frac{dx}{dt} = -x + y^2,$$

$$\frac{dy}{dt} = -y + x^2.$$

This system has two equilibrium points: $P = (0, 0)$ and $Q = (1, 1)$. The overall phase portrait of this system is shown schematically in Figure 2.26, where the boundary of the stable manifold for equilibrium point P is the curve marked A. This is a closed curve starting at $-\infty$ and going through the other equilibrium point Q. So all points inside the region bounded by curve A are attracted to the origin. For this reason, curve A is sometimes called the boundary of the *domain of attraction* of equilibrium point P at the origin.

The foregoing example has shown that the boundary of the domain of attraction of an equilibrium point can, and usually is, a simple closed curve (or, in higher dimensions, a hypersurface). A similar argument applies for the case of the other classical attractor, a limit cycle. But what about a strange attractor? What is the geometrical nature of the boundary of its domain of attraction? This is where fractals make their appearance in the theory of dynamical systems. But before developing that connection, let's first discuss what a fractal actually is.

A Mountain Is Not a Cone and a Cloud Is Not a Sphere

Experiencing the world ultimately comes down to the recognition of boundaries: self/nonself, past/future, inside/outside, subject/object, and so forth. And so it is in mathematics, too, where we are continually called upon to make distinctions: solvable/unsolvable, computable/uncomputable, linear/nonlinear, and other categorical distinctions involving the identification of boundaries. In particular, in geometry we characterize the boundaries of especially important figures by giving them names such as circles, triangles, ellipses, and spheres. But when it comes to using these kinds of boundaries to describe the natural world, these simple geometrical shapes fail us completely: mountains are not cones, clouds are not spheres, and rivers are not straight lines.

To illustrate the point, Figure 2.27 shows the coastline of Norway. It's patently obvious that the boundary of Norway is not any kind of simple geometrical curve, but rather is a very complicated, twisty, jagged kind of line. Moreover, even though the square grid in the figure is about $\delta = 50$ km on a side, the profile of the coastline doesn't get any smoother if we refine the scale. For example, taking $\delta = 10$ km (or even $\delta = 1$ km) on a side yields exactly the same kind of jagged boundary. To see this, note that if the coastline were a regular geometric shape such as a series of connected straight lines, then the overall length of Norway's border should remain the same, regardless of the length of the "yardstick" we

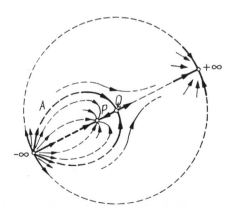

Figure 2.26. The domain of attraction for equilibrium point P at the origin.

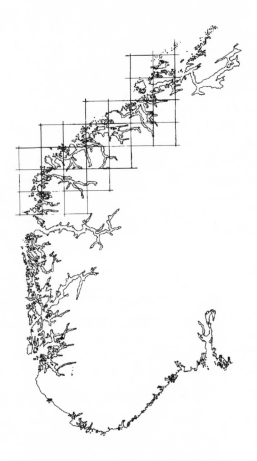

Figure 2.27. The coastline of Norway.

employ to measure it. But, as shown in Figure 2.28, when we refine our scale by reducing δ, the overall length of the border systematically increases. This is a surefire tip-off that there's something going on here that lies beyond the bounds of classical geometry. In this section we'll look at what this "something" is.

The kind of curve describing the coastline of Norway is what in 1975 IBM mathematician Benoit Mandelbrot labeled a *fractal*. Fractals are curves that are irregular all over. Moreover, they have exactly the same degree of irregularity at all scales of measurement. So it doesn't matter whether you look at a fractal from far away or close up with a microscope—in either case you'll see exactly the same picture. If you

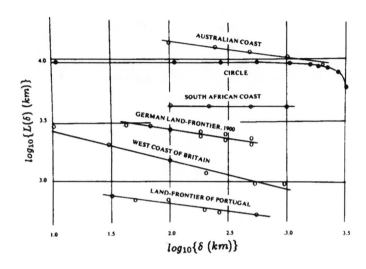

Figure 2.28. The measured length of the coastline at different scale lengths.

start looking from a distance, then as you get closer small pieces of the curve that looked like formless blobs before turn into well-defined objects, objects whose shape is the same as that of the whole object seen previously.

There are many examples of fractals in nature: ferns, clouds, lightning bolts, coastlines, river basin networks, and galaxies, to name but a few. We'll look at some of these later. For now, let's look at a simple example of a fractal pattern called the *Sierpinski gasket.* The basic idea underlying the construction of this strange object is illustrated in Figure 2.29. The gasket is formed by starting with the black equilateral triangle at the top. The white triangle is then cut out of the center, leaving three smaller black equilateral triangles. This excision process is then repeated on the three black triangles, obtaining nine new black triangles. This process is then repeated indefinitely, doubling the detail at each stage. The pattern obtained as this process is continued indefinitely is the famous Sierpinski gasket.

As the preceding examples show, for complicated geometrical objects the ordinary notion of dimension may vary with scale. For fractals, the counterparts of the familiar integer dimensions of, for example, points, lines, and squares are values that are not usually whole numbers. The simplest kind of fractal dimension is called the *similarity dimen-*

Figure 2.29. The Sierpinski gasket.

sion D. If we have a linearly self-similar curve, it could be anything from an almost smooth one-dimensional line to a nearly plane-filling curve that visits virtually every part of the plane. In the first case, D will be only slightly greater than one; in the second, D will be very nearly two. Intuitively speaking, the similarity dimension measures the degree of "roughness" of a fractal shape. Let's see how to calculate this quantity.

Computing the Fractal Dimension

If we have a line segment of unit length, then we can cover the line by N nonoverlapping lines of length $1/N$ for any positive integer N. We then say that the line is (linearly) self-similar with scaling ratio $r(N) = 1/N$. Similarly, a rectangle in the plane may be covered by scaled-down rectangles if we change the length scale by $r(N) = (1/N)^{\frac{1}{2}}$. For cubes, the appropriate scaling factor is $r(N) = (1/N)^{\frac{1}{3}}$, and for a general self-similar shape we use a scale factor $r(N) = (1/N)^{\frac{1}{D}}$, where D is the similarity dimension. If we rewrite this expression, we can obtain $D = -\log N/\log r(N)$. So, for example, in our construction of the Sierpinski gasket, the filled-in triangles are replaced by $N = 3$ triangles, each of which has been scaled down by a factor $r = 1/2$. Therefore, the similarity dimension of the Sierpinski gasket is $D = \log 3/\log 2 \approx 1.58$.

But how do we compute the similarity dimension in practical cases such as the Norwegian coastline shown in Figure 2.27? The procedure is rather straightforward. First, cover the region with a set of squares having edge length δ. Let $N(\delta)$ be the number of such squares needed to cover the coastline. It follows from the preceding discussion that in the limit of small δ, we must have $N(\delta) \approx 1/\delta^D$. Because on a log-log scale the relationship $N(\delta)$ versus δ is linear, we can determine the fractal dimension of the coastline by finding the slope of the quantity $\log N(\delta)$ viewed as a function of $\log \delta$. We find this straight line by computing $N(\delta)$ for smaller and smaller values of δ. Now let's use this same general idea in another context to calculate the similarity dimension of a river basin network.

The Fractal Dimension of a River Basin

Consider the drainage area A shown in Figure 2.30, where L denotes the length of the longest river in the network. It is clear from the figure that for rivers with a constant drainage width d, the relationship between L and A is simply $L = A/d$. In that case, the similarity dimension is $D = 2$. On the other hand, if the drainage basin kept the same shape when magnified, then we would obtain $D = 1$. In 1982, Mandelbrot suggested the following length-area relationship for real drainage networks:

$$L(\delta) = k\delta^{1-D}A(\delta)^{\frac{D}{2}},$$

where L is the length of the perimeter bounding the area A, δ is the "yardstick size" used to measure the length, and k is a proportionality constant. Note here that the area A of the region is also a function of δ, and is assumed to be measured by covering it with little squares having area δ^2.

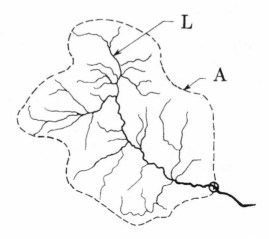

Figure 2.30. A river drainage area.

Studies of drainage networks in the northeastern United States using topographic maps and aerial photographs have shown the empirical relationship

$$L = 1.4A^{\frac{D}{2}},$$

with $D = 1.2$. Consequently, the fractal dimension of most rivers and streams is $D = 1.2$, and, as a result, there is good evidence for concluding that in this region the stream length is, on the average, proportional to the $D/2$, that is, the 3/5th, power of the drainage area—regardless of the geological or topographic features of the land area.

The preceding examples show that natural phenomena often involve something happening on all length scales. What seems to make many self-similar natural objects interesting is just this fact—that they possess significant features on many length scales. But spatial lengths are not the only intervals that we care about in life. Temporal lengths and frequency intervals are two others. Music is a good example of this because no matter on what scale the music is written, it must have pitch changes on many frequency scales and rhythm changes on more than one time scale in order to be aesthetically appealing. Let's illustrate this point by considering some of the music written by Johann Sebastian Bach and how it might be "downsized" by using these ideas of self-similarity and fractals.

Bach and Fractal Music

If an electrical engineer were to compute the power spectrum $f(x)$ of the relative frequency intervals x between successive notes in Bach's *Brandenburg Concerto,* over a large range of frequency intervals the relation $f(x) = c/x$ holds, where c is some constant. Thus, Bach's music is characterized by the kind of "noise" that engineers call *1/f noise.* This hyperbolic functional relationship between the frequency and the intervals between successive notes shows up in the spectrum of amplitudes, too, as seen in Figure 2.31, which shows the amplitude spectrum for Bach's *First Brandenburg Concerto.* The question, of course, is why Bach and so many other composers have created music that seems to obey this $1/f$ rule.

Part of the answer to this puzzle lies in the simple observation that for any piece of music to be "interesting," it should be like Goldilocks's porridge: not too hot, and not too cold—but just right. In power spectrum terms, this means that the relationship between the power and the frequency should not behave like boringly predictable "brown" noise, with its $1/f^2$ frequency dependence, or like completely unpredictable

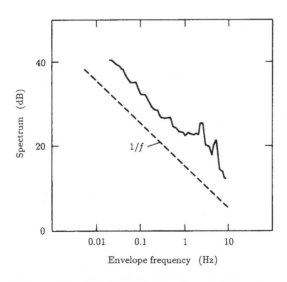

Figure 2.31. The amplitude spectrum for the *First Brandenburg Concerto.*

93

"white" noise and its $1/f^0$ dependence. Rather, it is the intermediate "pink" noise case that corresponds to our aesthetically satisfying $1/f$ relationship. Figure 2.32 shows examples of music composed according to each of these frequency patterns. Part (a) shows "white" music produced from independent notes, part (b) is "brown" music composed of notes with independent frequency increments, and part (c) consists of "pink" music whose notes have a frequency and duration determined by $1/f$ noise.

The homogeneous power laws obeyed by fractal objects suggests a way of "compressing" the music of a Bach or a Mozart down to its irreducible "essence." This idea, in turn, opens up the possibility of perhaps using this distilled Bach essence to compose new "Bach-like" music. Physicist Kenneth Hsu took up this challenge. Figure 2.33 shows

Figure 2.32. (a) White, (b) brown, and (c) pink music.

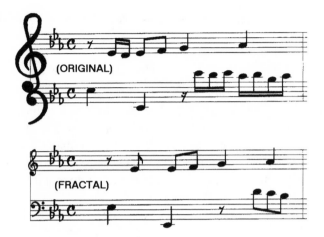

Figure 2.33. Bach's *Invention 5* in its original and fractal forms.

the results of some of his work. In the figure, the notes are plotted using the frequency and amplitude spectra of Bach's music to determine each note's relationship to its neighbors. By removing notes from several of Bach's inventions, Hsu found that basic patterns persisted in the fractal reductions of the music, even if what remained contained as little as 1/64th of the original notes. Thus, music recognizable as Bach survives fractal reduction. And, in fact, to some ears this "reduced Bach" gives the impression of an economy of frills and ornamentation. Figure 2.33 shows one of these reductions in the case of Bach's *Invention 5*. With these preliminaries about fractals in hand, let's now return to our original goal of establishing the connection between fractals and the boundaries of the domains of attraction for strange attractors.

Fractals and Domains of Attraction

We have seen above that the most important thing one can know about a dynamical system is its set of attractors. Associated with each attractor, there is a set of initial states that winds up in that attractor under the action of the system's vector field. These points form the domain of attraction of the attractor. Here we show that even for the simplest kind of nonlinear dynamics, the structure of the domain of attraction often has a fractal boundary.

Fix a point c in the complex plane. Now form a sequence of complex numbers by the rule

$$z_k = z_{k-1}^2 + c, \qquad z_0 = 0, \qquad k = 1, 2, \ldots . \qquad (\P)$$

So for each choice of parameter c, we have a different quadratically nonlinear dynamical system. It's clear that for some points c, the sequence $\{z_k\}$ will diverge to infinity. The choice $c = 2$ is such a point. But for other points, such as $c = i$, the sequence remains bounded. We now form a set M in the complex plane in the following way: We put the number c in M if the sequence $\{z_k\}$ does *not* go off to infinity; otherwise, c is not in M. So, for instance, the point $c = 2$ is not in M, whereas the point $c = i$ is. The set M is called the *Mandelbrot set* and is shown in Figure 2.34.

It turns out that when the starting point of our iteration is in the interior of M, the quadratic map ($\P$) yields an orbit that is orderly and

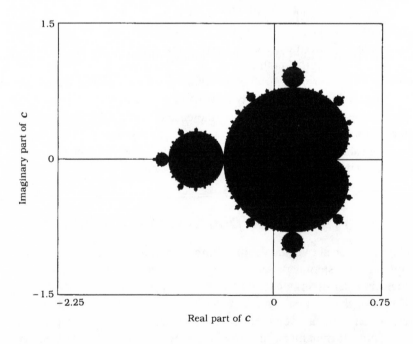

Figure 2.34. The Mandelbrot set.

96

well-behaved. But if z_0 is outside M, the orbit, while perfectly deterministic, is wild and disorderly, wandering all over the complex plane. The boundary of M, separating the orderly and disorderly behaviors, turns out to be unbelievably messy. As one might surmise from inspecting Figure 2.34, new copies of the entire Mandelbrot set continue to "bud off" from the original at all scales. Furthermore, there are many additional structures that appear at various magnification levels. For instance, Figure 2.35 shows a magnified view of part of the boundary of the set M. New structures continue to appear as one goes to higher and higher magnifications because the set M remains forever intricate on any scale. This kind of never-ending sequence of new patterns has led many to claim that the Mandelbrot set is the most complicated object known to humankind.

To add further ammunition to this claim, in 1991 the Japanese mathematician Mitsuhiro Shishikura proved that the dimension of the boundary of the Mandelbrot set is 2. This means that not only is the boundary of M indeed a fractal, but that it "wiggles" as much as any curve in the plane possibly can. In short, if you measure the complexity of a curve by its "wiggliness," then no curve can be more complex than the boundary of M. However, this does *not* mean that the boundary of M is a space-filling curve. Merely having dimension 2 is not enough for this, as is shown by the case of a two-dimensional sponge, which is

Figure 2.35.   Magnified view of part of the boundary of the Mandelbrot set.

all "holes." Whether the boundary of the Mandelbrot set fills up a two-dimensional continuum in the plane is, at present, still an open question.

The Mandelbrot set can be thought of as characterizing the domain of attraction of the starting point $z_0 = 0$ for the *family* of quadratic maps $z \rightarrow z^2 + c$. But this is not usually what we mean when we speak of the domain of attraction. The usual idea is that we have a *single* system with one or more attractors, and we want to characterize those initial states that lead to the different attractors. It turns out that if the system has a strange attractor, the boundary of that kind of attractor has remarkable fractal properties.

To see this, let's go back to our earlier quadratic mapping

$$z_k = z_{k-1}^2 + c, \qquad k = 1, 2, \ldots,$$

and fix the parameter value at $c = -1$. Now we consider the set of initial points z_0 such that the sequence $\{z_n\}$ remains bounded as $k \rightarrow \infty$. The boundary of the black part of Figure 2.36 shows these points, and the black interior is the strange attractor for this system. The set of boundary points is called the *Julia set* of the system, and the filled-in black part is called the *filled-in Julia set*. For the sake of completeness, let's note that the complement of the Julia set (the white part) is called the *Fatou set*. These names owe their origin to the French mathematicians Gaston Julia and Pierre Fatou, who studied this kind of set in the early part of the twentieth century.

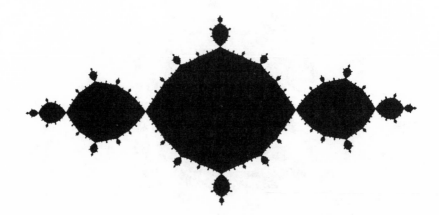

Figure 2.36. The Julia set of the quadratic map $z \rightarrow z^2 - 1$.

It is interesting to see what happens to the Julia set for the quadratic map when $c = 0$. In that case, the iterates are just the powers of z, z^2, z^4, z^8, and so on. Clearly, if the initial point z_0 lies inside the unit circle, then its orbit remains bounded; if it starts outside the unit circle, it flies off to infinity. And if $|z_0| = 1$, the orbit moves around the circle indefinitely. Therefore, for this system the Julia set is simply the unit circle $|z| = 1$. Amazing things now happen to the Julia set when we perturb c away from zero.

For the slightest nonzero value of c, the Julia set gets distorted—but not into a smooth, roughly circular curve. No. What happens is that the curve becomes a fractal. Figure 2.37 shows a sequence of distortions of the original unit circle as we gradually increase the magnitude of c, going from the "almost circle" in part (a) to the so-called Douady rabbit in part (b), and on to a set of disconnected islands in part (c).

This quadratic map, elementary as it is, shows us that the boundary of the domain of attraction of a strange attractor can be just about as complicated as a mathematical object can get. And, in fact, it's rather easy to see that the Julia set for this map and the Mandelbrot set are intimately related. But to get a real feel for the depth and beauty of these objects, the more detailed literature cited in the references is must reading.

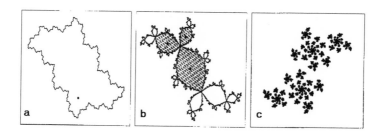

Figure 2.37. The Julia set for various nonzero values of c.

CHAPTER

3

The Kalman Filter

Control Theory

Looking for Life

In summer 1975, two Viking spacecraft were launched from Cape Canaveral, Florida, on their way to a rendevous with Mars one year later, where their mission was to search for signs of life on the Martian surface. Upon arrival, both spacecraft were put into a predetermined orbit around Mars, from which the search for suitable landing places began. On July 20, 1976, the *Viking I* lander came to rest in the Chryse Planitia region of Mars, some 23 degrees north of the equator, and six weeks later *Viking II* settled down onto the Utopia Planitia plain, which is almost exactly 180 degrees of longitude away from the first landing site, thus putting the two landers on opposite sides of the planet.

There were six different instruments on the landers used for the detection of signs of life: two cameras for photographing the landscape, a combined gas chromatograph and mass spectrometer for analyzing the surface for organic material, and three instruments designed to detect the metabolic activities of any microorganisms present in the Martian soil. Sad to say, none of the tests produced unequivocal evidence of life at either of the landing sites. Nevertheless, much data of great scientific value was obtained, including signs of extraordinary chemical reactivity of the Martian soil. But for our story here, what is notable about the entire Viking mission is that it took place at all! Think for a moment what was involved in just getting the two landers to their antipodal locations on the Martian surface.

Leaving aside the mechanical, electrical, and chemical engineering aspects of the Viking spacecraft, the sine qua non of the mission is the obvious point that NASA had to be able to get the landers from Cape Canaveral to their resting points on Mars. This task involved several aspects of guidance and control: determining what kind of propulsion was necessary to transfer a certain kind of vehicle from Earth to a Martian orbit; calculating the various trajectories possible; identifying the trajectories of minimal cost in some combination of time, energy, and other resources; developing the capability to measure where the spacecraft was located at any time during its flight; and being able to exert midcourse corrections if and when the spacecraft drifted off its design trajectory. At the center of all these tasks sits what has come to be called the *Kalman filter,* a mathematical theory that, along with the simplex method of optimization theory (see Chapter 5 of *Five Golden Rules*), is probably the single most *useful* piece of mathematics developed in this century. Our story in this chapter is an account of what this gadget is and how it came to occupy its pivotal role in the mathematical theory of control systems.

Can You Get There from Here?

To see what's involved in the problem of guidance and control, let's look at a simple toy version of the problem. Suppose we have a spacecraft whose trajectory is constrained to move along the real line, as shown in Figure 3.1. If the vehicle starts at the origin with an initial velocity in the positive direction, then it's clear that by simple inertia it will just sail along, eventually visiting every location on the positive real axis. But if we want to bring the spacecraft to a soft landing at, say, the point marked X in Figure 3.1, then we would have to turn it around somehow, and then exert a retrothrust to neutralize the inertia from the rocket's initial thrust, thereby reducing the vehicle's velocity to zero at the point X. Under what circumstances can this be done? In other words, when can we transfer the spacecraft from the origin with a given initial velocity to the point X with zero velocity? This is a question that control theorists call a problem of *reachability.*

To see what's at issue in solving the reachability problem, consider a linear, discrete-time control system moving in the plane. To make things as transparent as possible, suppose the control input at each moment is a simple scalar, and that it enters into the dynamics in a linear fashion. Under these conditions, the system dynamics can be written in the form

$$x_1(t + 1) = f_{11}x_1(t) + f_{12}x_2(t) + g_1u(t),$$
$$x_2(t + 1) = f_{21}x_1(t) + f_{22}x_2(t) + g_2u(t). \tag{$*$}$$

$$O \qquad\qquad\qquad\qquad\qquad\qquad X$$

Figure 3.1. A spacecraft moving along the real line.

For ease of writing, let's write the above system in the vector-matrix form

$$x(t + 1) = Fx(t) + gu(t),$$

where F and g are given by

$$F = \begin{pmatrix} f_{11} & f_{12} \\ f_{21} & f_{22} \end{pmatrix}, \qquad g = \begin{pmatrix} g_1 \\ g_2 \end{pmatrix}.$$

Here the quantities $f_{ij}, g_i (i, j = 1, 2)$ are just real numbers. Furthermore, it turns out to involve no loss of generality to assume that the initial state of the system is $x_1(0) = x_2(0) = 0$, that is, the system starts at the origin. The reachability problem can now be stated as: What points (x_1, x_2) in the plane can be "reached" within, say, T time steps by a suitable choice of the control inputs $u(0), u(1), \ldots, u(T - 1)$?

First of all, let's dispose of the two extreme cases, (1) when there is no control at all ($u(t) = 0$ for all t), and (2) when the control input cannot influence the system state (if $g_1 = g_2 = 0$). In both these cases, the system state remains at the origin forever; hence, no state other than the origin is ever reachable. Trivial as these two cases are, they illustrate the twin points that (1) a control system isn't a control system if you can't exert some influence on the behavior of the system's state, and (2) the only way you can influence the state is when the channel between the control and the system state remains open. So we will henceforth assume without further comment that these conditions are satisfied.

Suppose, then, that the initial control input $u(0)$ is nonzero. Then the system state at time $t = 1$ is $x_1(1) = u(0)g_1$ and $x_2(1) = u(0)g_2$. Thus, as long as the quantities g_1 and g_2 are not both zero, the system moves away from the origin after one time step to the point $(g_1, g_2)u(0)$. If we introduce the state vector $x(t) = (x_1(t), x_2(t))'$ and the input vector $g = (g_1, g_2)'$, this can be written more compactly as $x(1) = gu(0)$. (*Note:* Here the notation v' means simply the transpose of the vector v.) This means that in one time step the system can be transferred to any point on the line from the origin through the point g. By a similar line of argument, the state at the second time step is

$$\begin{aligned} x(2) &= (f_{11}x_1(1) + f_{12}x_2(1) + g_1u(1), f_{21}x_1(1) + f_{22}x_2(1) + g_2u(1))' \\ &= (u(0)[f_{11}g_1 + f_{12}g_2] + g_1u(1), u(0)[f_{21}g_1 + f_{22}g_2] + g_2u(1))' \\ &= Fgu(0) + gu(1). \end{aligned}$$

Geometrically, we represent these state transitions in Figure 3.2. These calculations show that by a suitable choice of the control inputs $u(0)$ and $u(1)$, it is possible to transfer the system state from the origin to any point that can be written as a linear combination of the two vectors g and Fg. So if these vectors are linearly independent, there is some sequence of control inputs transferring the system state from the origin to *any* point in the (x_1, x_2)-plane. We call such a control system *completely reachable*. Geometrically, these results say that any point that can be reached by application of a control input lies in the subspace generated by the vectors g and Fg.

To summarize, the system is completely reachable if and only if the vectors g and Fg are linearly independent (i.e., they generate the entire state space). Moreover, in that case any state can be reached from the origin in not more than two time steps. Now let's look at the general situation.

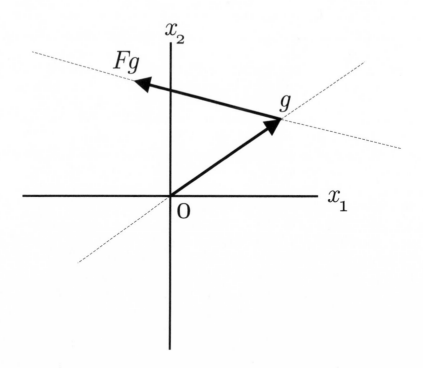

Figure 3.2. The first two state transitions for the system ($*$).

The Problem of Reachability

The reachability results obtained above impose several restrictions on the control system and the admissible control inputs. We now want to examine these assumptions explicitly, looking for how they can be relaxed and still retain the basic conclusions. The main points that need addressing are the following:

- *Linearity:* The system dynamics (∗) are linear with constant coefficients (the matrix F and the vector g are have constant entries), while the control input enters additively into the dynamics.
- *Unbounded, scalar control:* No constraints have been imposed on the magnitude of the control inputs; the quantities $u(0), u(1), \ldots$ can be any sequence of real numbers.
- *Discrete time set:* We have assumed that the system changes state only at integer moments in time, that is, the time set of the system is the nonnegative integers $t = 0, 1, 2, \ldots$.
- *Planar system:* Our results have been obtained only for a two-dimensional system whose state space is the entire plane R^2.

Happily, it turns out that the reachability conditions can be generalized to accommodate nonlinear processes, constraints on the control, and continuous time and state spaces of arbitrary, even infinite, dimension. Let's briefly consider these matters in the order opposite to the way things are listed above.

Suppose we have a discrete-time n-dimensional linear control system with m control inputs. This means that the system state is a vector in R^n, whereas the control inputs are now vectors in R^m. Mathematically, we can write such a system in the form

$$x(t + 1) = Fx(t) + Gu(t), \qquad (\ast\ast)$$

where F and G are $n \times n$ and $n \times m$ real matrices, respectively, and $x(t)$ is in R^n and $u(t)$ is in R^m. Note that here we still retain the assumptions of discrete time and unbounded control. We can generalize the earlier reachability result for the system (∗∗) as follows:

REACHABILITY THEOREM   *The linear dynamical system (∗∗) is completely reachable if and only if the block matrix*

$$\mathcal{C} = \left[G|FG|F^2G|\cdots|F^{n-1}G\right]$$

contains n *linearly independent vectors, that is, rank* $\mathcal{C} = $ n.   ∎

The proof of this fundamental result follows exactly the same lines as for the planar case given earlier. After the initial step, the system state is $x(1) = Gu(0)$. This means that any state that is a linear combination of the columns of G is reachable in one time step. Similarly, at the second time step we have $x(2) = FGu(0) + Gu(1)$, implying that any state that is a linear combination of the columns of the matrices FG and G can be reached in two time steps. Carrying on in this way, we arrive at the Reachability Theorem. (*Note:* Just in case you're wondering why we stop after at most $n - 1$ steps, the reason is that the Cayley-Hamilton Theorem of elementary matrix theory implies that for an $n \times n$ matrix A, the columns of all powers of A higher than $n - 1$ are linear combinations of the columns of the matrices $I, A, A^2, \ldots, A^{n-1}$. In system-theoretic terms, this means that if you haven't been able to reach a state by $n - 1$ time steps, you'll never get there—it can't be done.)

So we see that there is no problem whatsoever in extending the planar reachability results to the case of n-dimensional systems. The same holds true when we consider continuous rather than discrete time systems. So if the control system is given by the continuous time, n-dimensional process

$$\frac{dx}{dt} = Fx(t) + Gu(t),$$

where the matrices F and G are as before, then the condition for the system to be completely reachable is *exactly* the same as that given in the Reachability Theorem, namely, that the reachability matrix $\mathcal{C}$ be of full rank (i.e., contain n linearly independent vectors). In fact, for continuous time systems, if a state can be reached at all, then it can be reached in an arbitrarily short amount of time. This follows from the fact that we are imposing no finite bound on the magnitude of the control inputs. This will certainly raise the eyebrows of a hard-nosed, practically minded control engineer. But it's theory we're talking about here, not practice, and the mathematics says that if you can pump an arbitrarily large amount of energy into a system in a short enough period of time, you can do almost anything—provided the energy can actually reach the system state.

Now let's illustrate these reachability results with a quasi–real-life example from economics.

National Economic Planning

Just about every country in the world is faced with trying to juggle the following two economic objectives: (1) achieving full employment without inflation (internal economic balance), and (2) maintaining a positive, or at least neutral, balance of international payments (external economic balance). These goals have to be achieved through economic policy instruments such as changes in interest rates and in the budget deficit. As a result, the country has a central bank that controls interest rates and a legislative body that controls changes in the government's budget deficit. Although it is politically difficult to combine these institutions into a single controlling agency, it is possible to establish general directives for both of them to follow. Our question here concerns whether the actions of the central bank and the congress taken together can achieve the country's overall economic goals.

To study this problem, let's define the following variables:

$Y(t)$ domestic production (i.e., the income of consumers)
$X(t)$ national expenditures
$C(t)$ national consumption
$S(t)$ national savings
$I(t)$ domestic investments
$M(t)$ imports of foreign goods and services
$K(t)$ net capital outflow
$T(t)$ net taxes on transfers
$A(t)$ government expenditures on goods and services
$B(t)$ international balance-of-payments surplus
$r(t)$ domestic interest rate

All of these variables are annual rates in time period t and are deflated to a uniform price level. Let us define also the following variables, which are all assumed to be constant:

E exports of goods and services

Y_F domestic production with full employment and no inflation

r_F foreign interest rate

Four accounting identities link these variables:

$$Y = C + S + T, \qquad X = C + I + K + A,$$
$$B = E - M - K, \qquad B = Y - X.$$

Furthermore, the following linear relationships have been found to be approximately valid:

$$S = \alpha_1 Y - \alpha_0, \quad M = \beta_1 Y - \beta_0, \quad I = \gamma_1 r + \gamma_0, \quad K = \delta_1 r + \delta_0,$$

where the quantities α_i, β_i, γ_i, and $\delta_i (i = 0, 1)$ are empirically determined constants of proportionality.

If we define the state and control vectors as

$$x(t) = \begin{pmatrix} B(t) \\ Y(t) \end{pmatrix}, \qquad u(t) = \begin{pmatrix} r(t+1) - r(t) \\ D(t+1) - D(t) \end{pmatrix},$$

where $D(t) = A(t) - T(t)$ is the net government deficit, the above relations lead to the linear control system

$$x(t+1) = x(t) + Gu(t),$$

where

$$G = \begin{pmatrix} \delta_1 + \mu \beta_1 \gamma_1 & -\mu \beta_1 \\ -\mu \gamma_1 & \mu \end{pmatrix}, \qquad \mu = 1/(\alpha_1 + \beta_1).$$

From a given initial state $x(0)$, the government's economic planners want to steer this economy to the target state

$$x^* = \begin{pmatrix} 0 \\ Y_F \end{pmatrix},$$

in which the international payments are in balance and an internal level Y_F of outputs is attained that involves full employment and no inflation.

For this economy, it doesn't take too much thought to see that if there are no restrictions on the changes in interest rates and deficits, any desired state of the economy can be reached if the matrix G is nonsingular. Because this will be the case for almost every choice of the quantities δ_1, β_1, γ_1, and α_1, the complete reachability of this economy is "stable" with respect to small changes in these parameters.

The foregoing economic example shows how unrealistic it is, in general, to assume that an unlimited amount of influence can be brought to bear on a system. Control costs *something*, either in time, men, money, energy, or some other currency appropriate for the system at hand. So let's see what can be salvaged from these reachability results when restrictions are imposed on the control inputs.

There are many ways in which control actions can be constrained, so here we just give a flavor of the way in which reachability results are modified by constraints on the control inputs by considering the situation in which the inputs are restricted to be nonnegative. This is probably the most common type of control constraint for the simple reason that many controlling actions make no sense unless the input is positive or zero. For example, if the task is to detoxify a river or lake by putting in chemical nutrients, it's difficult to imagine what it would mean to put in a negative amount of nutrient. A similar situation arises when, for example, we wish to regulate a person's heartbeat by administering medication. Here we can't give the person a negative amount of medicine. For discussions of other types of control constraints, the reader is referred to the material cited in the chapter references.

Suppose we have the scalar control system

$$\frac{dx}{dt} = u(t).$$

From our earlier results, it is clear that this system is completely reachable. But if we impose the condition that $u(t) \geq 0$ for all t, then the system is not positively reachable; the only states that can be reached by such a system are those lying on the positive real axis. Similarly, the planar system

$$\frac{d^2x}{dt^2} = u(t),$$

is also not positively reachable because for this system the only states that can be reached from the origin are those of the form

$$x = a_0 g + a_1 F g, \qquad a_0, a_1 > 0,$$

where

$$g = \begin{pmatrix} 0 \\ 1 \end{pmatrix} \qquad \text{and} \qquad Fg = \begin{pmatrix} 1 \\ 0 \end{pmatrix}.$$

In other words, only points in which both $x(t)$ and dx/dt are positive are reachable from the origin using positive controls. Geometrically, these are the points in the first quadrant of the $(x, dx/dt)$ phase plane. Finally, consider the system

$$\frac{dx_1}{dt} = x_1 \cos \theta + x_2 \sin \theta,$$

$$\frac{dx_2}{dt} = -x_1 \sin \theta + x_2 \cos \theta + u(t), \qquad 0 < \theta \neq k\pi, \quad k = 1, 2, \ldots .$$

This system is completely reachable using positive controls $u(t) > 0$. Intuitively, the reason a positive control can drive the system to any state in the plane is that we can use the positive control to go out any distance we wish from the origin—but not in any direction. Once the system has been driven a distance equal to that of the state we want to reach, however, the control can then be turned off and we can then let the natural oscillatory dynamics of the system carry the state around to the desired end state. This situation is depicted in Figure 3.3.

From these intuitive arguments, two key points emerge regarding positive reachability for a scalar-input system. First of all, the system's uncontrolled, or "free," dynamics must cause the system trajectory to visit all orthants of the state space, that is, the positive and negative directions along each coordinate axis. Second, it may take some amount of time to reach a given state. Of course, there is also the obvious requirement that for the system to be positively reachable it must be reachable in the usual sense. By putting all these remarks together, we have the Positive Reachability Theorem.

POSITIVE REACHABILITY THEOREM   *The single-input system*

$$\frac{dx}{dt} = Fx + gu(t),$$

is completely reachable with positive inputs u(t) $\geq$ 0 *if and only if*

1. *the system is completely reachable for unbounded inputs, that is, the reachability matrix C is of rank* n;

2. *the eigenvalues of the matrix F all lie on the imaginary axis;*

3. *the time* T *when a state is to be reached is large enough.* ∎

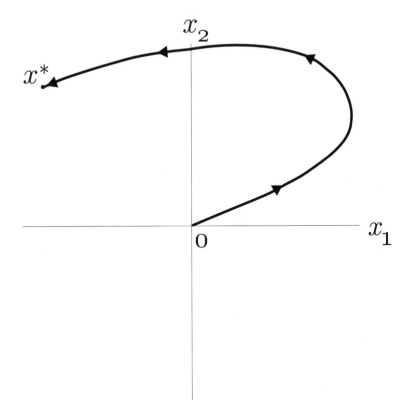

Figure 3.3. Positive reachability.

Let us note in passing that this result implies that no single-input system of odd order can be positively reachable because there must then be a real eigenvalue of the matrix F (unless that eigenvalue happens to be 0). The overall situation eases somewhat in the case of multiple inputs, but this is a matter beyond the scope of the elementary account being given here.

So far, we have addressed the problem of high-dimensional systems, multiple inputs, and control constraints. This leaves the matter of linearity. How dependent are the foregoing reachability results on the linear nature of the dynamics and the fact that the control inputs enter the dynamics additively? Basically, the answer is: Not much. But before taking up the reasons why, it's worth dallying a few moments

113

longer in the linear realm to examine what amounts to the other side of the reachability coin, the problem of observability.

The Problem of Observation

The results of many medical tests such as blood sugar content, urinalysis, and body temperature are used as indirect indicators of more basic bodily processes that we want to know about but cannot, either because it's too dangerous or just plain impossible to measure the quantities directly. Medicine is not the only realm, however, where such inferences about the actual internal state of the system have to be made. For example, your car's oil and water temperature gauges are merely indicators of far more basic—and important—processes going on inside the car's engine. The theory of system *observability* addresses the degree to which we can infer a system's true internal state from these kinds of indirect measurements.

The observability question is particularly well illustrated by the problem mentioned a moment ago of determining the concentration of a drug in a patient's body by measurements on, say, the concentration of the drug found in the urine. For example, think of a cardiac patient who receives the drug digitoxin, and who then metabolizes it to digoxin. There is a rather fine line between receiving a lethal dose of digitoxin and receiving the amount necessary for the drug to have a therapeutic effect, so it is of great importance to know accurately the amount present in the body before prescribing additional doses.

A multicompartment model used to describe the kinetics and metabolism of digitoxin is shown in Figure 3.4, where X represents the amount of digitoxin in the body, Y is the amount of digoxin, and S_1 and S_2 are urinary excretion sinks, say, from the kidneys to the bladder and from the bladder to the outside. The quantities k_1 through k_5 are constants characterizing the rate of diffusion between the various compartments.

In the medical community, it is more or less standard practice to assume that when a dose D of digitoxin is given, approximately 92 percent of the dose is immediately taken up by compartment X, and about 85 percent of the remaining 8 percent is instantly taken up by compartment Y. With these assumptions, together with the compartmental

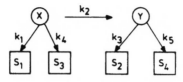

Figure 3.4. Multicompartment structure for digitoxin metabolism in the body.

structure shown in Figure 3.4, we can write down the dynamics of the drug concentrations as

$$\frac{dX}{dt} = -(k_1 + k_2 + k_4)X, \qquad \frac{dY}{dt} = k_2X - (k_3 + k_5)Y,$$

$$\frac{dS_1}{dt} = k_1X, \qquad \frac{dS_2}{dt} = k_3Y, \qquad \frac{dS_3}{dt} = k_4X, \qquad \frac{dS_4}{dt} = k_5Y.$$

The initial conditions are

$$X(0) = 0.92D, \qquad Y(0) = (0.85)(0.08)D,$$
$$S_1(0) = S_2(0) = S_3(0) = S_4(0) = 0.$$

What we want to know is the unknown initial dosage D because this will tell us how much medication is in the patient's body at any future time. But we have at our disposal only measurements from the two urinary excretion sinks $S_1(t)$ and $S_2(t)$ over some time interval, $0 \le t \le T$. So the observability question becomes: Given the measurements of $S_1(t)$ and $S_2(t)$ over the period $0 \le t \le T$, is it possible to pin down exactly the initial state $X(0)$ (or $Y(0)$) and, hence, the initial dosage D of digitoxin?

On the basis of the results to follow, we will see that it is *not* possible to determine the initial state of the system using urinary excretion measurements alone. Note, however, that this does not necessarily imply that D cannot be identified. It simply means that we cannot reconstruct the *entire* initial state of the system from knowledge of the functions $S_1(t)$ and $S_2(t)$. It still may be possible to determine that part of the initial state involving either X or Y. We shall see.

Complete Observability

Let's look again at the continuous time, linear control system

$$\frac{dx}{dt} = Fx(t) + Gu(t), \qquad x(0) = x_0, \qquad (\dagger)$$

where the observed output is given by

$$y(t) = Hx(t). \qquad (\ddagger)$$

Here $x(t)$ is an n-dimensional state vector and the output $y(t)$ is a p-dimensional vector, so that H is a $p \times m$ matrix. Upon integrating the state dynamics, we obtain

$$x(t) = e^{Ft}x_0 + \int_0^t e^{F(t-s)} Gu(s)\, ds,$$

so that the output

$$y(t) = He^{Ft}x_0 + \int_0^t He^{F(t-s)} Gu(s)\, ds.$$

This shows that the effect of the initial state on the output is decoupled from the effect of the control input. Therefore, if our goal is to identify the initial state x_0 from the measured output, there is no loss in generality in assuming that the control input $u(t) = 0$ for all time. (*Note:* This separation does not work, in general, when the system dynamics are nonlinear. In that case we have to explicitly account for the control input when assessing the system's observability.)

With the foregoing remarks in mind, the problem of observability reduces to the following: Can the initial state x_0 be uniquely identified on the basis of the output $y(t)$ measured over an interval $0 \le t \le T$? If so, then we call the system *completely observable*. The answer to the question is contained in the Observability Theorem.

OBSERVABILITY THEOREM   *The system $(\dagger)$–$(\ddagger)$ is completely observable if and only if the matrix*

$$\mathcal{O} = \left[H' | F'H' | (F')^2 H' | \cdots | (F')^{n-1} H' \right]$$

has n linearly independent rows, that is, $\mathcal{O}$ is of full rank. ∎

As an exercise, the reader might want to use the Observability Theorem to confirm the statement made earlier that the pharmacokinetics system for heart drugs is not completely observable. Now let's look at a different kind of example to illustrate this result.

Satellite Observations

The trajectory of a particle such as a communications satellite in near circular planetary orbit can be closely approximated by a linear system $dx/dt = Fx(t)$, where the state consists of both the radial and angular positions and velocities of the satellite. So the state vector $x(t)$ is in R^4. For such a system, the matrix F has the form

$$F = \begin{pmatrix} 0 & 1 & 0 & 0 \\ 3\omega^2 & 0 & 0 & 2\omega \\ 0 & 0 & 0 & 1 \\ 0 & -2\omega & 0 & 0 \end{pmatrix},$$

where ω is the angular velocity of the satellite when it moves in a completely circular orbit.

Suppose we can measure both the radial position $x_1(t)$ and the angular position $x_3(t)$. Thus, the matrix singling out these components of the state is

$$H = \begin{pmatrix} 1 & 0 & 0 & 0 \\ 0 & 0 & 1 & 0 \end{pmatrix}.$$

Appealing to the Observability Theorem, we calculate the observability matrix $\mathcal{O}$ to obtain

$$\mathcal{O} = \begin{pmatrix} 1 & 0 & 0 & 0 & 3\omega^2 & 0 & 0 & -6\omega^3 \\ 0 & 0 & 1 & 0 & 0 & -2\omega & -\omega^2 & 0 \\ 0 & 1 & 0 & 0 & 0 & 0 & 0 & 0 \\ 0 & 0 & 0 & 1 & 2\omega & 0 & 0 & -4\omega^2 \end{pmatrix}.$$

It's easy to see by inspection or otherwise that this matrix has four linearly independent rows; hence, it is of full rank. Thus, the satellite system is completely observable on the basis of measurements of both the radial and angular positions. This means that by position measurements alone, it is possible to pin down precisely the satellite's initial positions *and* velocities.

We now know that two sensors measuring radial and angular positions suffice to determine all aspects of the satellite's motion, so we might consider not measuring the angular position in an attempt to minimize the cost of making two measurements. We therefore ask: Are the satellite's position and velocity still determinable if we can measure only the radial position? Mathematically, this question translates into use of the matrix $H = (1 \quad 0 \quad 0 \quad 0)$ to single out the first component of the state as our sole observed quantity.

Using the foregoing matrix H, we again calculate the observation matrix $\mathcal{O}$ to obtain

$$\mathcal{O} = \begin{pmatrix} 1 & 0 & 3\omega^2 & 0 \\ 0 & 1 & 0 & -\omega^2 \\ 0 & 0 & 0 & 0 \\ 0 & 0 & 2\omega & 0 \end{pmatrix}.$$

A glance at the third row of this matrix leads immediately to the conclusion that $\mathcal{O}$ is not of full rank. So the system is not completely observable when we eliminate the measurement of angular position. On the other hand, a similar calculation shows that if we had eliminated measurement of the radial rather than the angular position, the system would have remained completely observable. We leave verification of this fact to the reader.

Duality

Comparison of the Reachability and Observability Theorems shows a striking similarity between the key objects $\mathcal{C}$ and $\mathcal{O}$ by which we test a system for complete reachability and complete observability, respectively. In fact, by making the substitutions

$$F \to F', \qquad G \to H',$$

we can transform the reachability matrix $\mathcal{C}$ into the observability matrix $\mathcal{O}$. This is an example of the powerful mathematical concept of *duality*. In this control system setting, duality enables us to translate any result about reachability and control inputs into an equivalent result about observability and measured system outputs. For instance, by duality we can show that the control system

$$\frac{dx}{dt} = Fx(t) + Gu(t)$$

is completely reachable if and only if the dual system

$$\frac{dx}{dt} = -F'x(t), \qquad y(t) = G'x(t)$$

is completely observable. So duality always enables us to get two theorems for the price of one.

It is important to note here how formation of the dual system not only involves using the transposed matrices F' and G', but also necessitates reversing the flow of time. This is a natural requirement because the normal direction of time involves a flow from the control inputs to the system outputs. So if we want to, in effect, interchange the roles of the system's inputs and outputs, we must also reverse the flow of time to respect this interchange. We'll encounter duality again in a later section of this chapter, where we will uncover an even more astonishing result: the equivalence between a completely deterministic control problem and a random process associated with observation.

Up to now, our focus has been on the inputs and outputs of linear systems. This is for several reasons. First of all, the questions of reachability and observability take on a particularly simple form in this setting, requiring only elementary linear algebra and matrix theory for their complete expression. Also, for linear systems we can easily separate the control inputs from the observations, as was noted earlier. This fact not only simplifies the analysis, but also enables us to give global results about observability without having to tie them to some particular class of control inputs. Finally, the linear case serves as a guide to what kind of results to look for when we leave the cozy confines of the linear world. All this having been said, let's briefly venture into the jungleland of nonlinearity to give at least a flavor of what can be done by way of extending the Reachability and Observability Theorems.

A Nonlinear Interlude

Think of a child pumping a swing, as depicted in Figure 3.5. Suppose we ask: Can the child pump the swing so as to make it reach any height she or he wishes? This is a problem of *nonlinear* reachability. To see why, suppose we model the situation by considering a mass m that moves up and down along a weightless rod (the child changing its center of mass). The state space for the problem is $M = \{(\theta, d\theta/dt, \ell):$

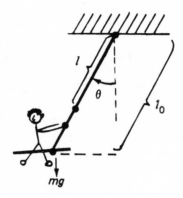

Figure 3.5. A child pumping a swing.

$0 \leq \theta \leq 2\pi, d\theta/dt \in R, 0 < \ell < \ell_0\}$. Setting the time rate of change of the angular momentum about the support equal to the torque due to gravity yields the dynamical equation of motion

$$\frac{d}{dt}\left(\frac{d\theta(t)}{dt}\ell^2(t)m\right) + \ell(t)m\sin\theta(t) = 0.$$

Differentiating this expression leads to the system

$$\frac{d^2\theta}{dt^2} + \left(\frac{2}{\ell}\frac{d\ell}{dt}\right)\frac{d\theta}{dt} + (1/\ell)\sin\theta(t) = 0.$$

If we now let $d\theta/dt = \phi$ and regard $d\ell/dt = u$ as the control, we obtain the system of three equations

$$\frac{d\theta}{dt} = \phi, \qquad \frac{d\phi}{dt} = -(1/\ell)[2u\phi + \sin\theta], \qquad \frac{d\ell}{dt} = u.$$

For later reference, it is convenient to write this system in the following vector form

$$\frac{d}{dt}\begin{pmatrix}\theta \\ \phi \\ \ell\end{pmatrix} = \begin{pmatrix}\phi \\ -(1/\ell)\sin\theta \\ 0\end{pmatrix} + u\begin{pmatrix}0 \\ -(2/\ell)\phi \\ 1\end{pmatrix} \doteq p(x) + u(t)g(x),$$

where we have defined the state vector x to be $x = (\theta, \phi, \ell)'$.

To see whether the child can pump the swing to a given amplitude, let's first consider the following more general situation. Suppose we have the nonlinear control system

$$\frac{dx}{dt} = \sum_{i=1}^{m} u_i(t) f_i(x(t)), \qquad x(0) = x_0 \in R^n.$$

For simplicity, we assume here that the functions $f_i(x)$ have whatever analytical properties we need for any particular operations we want to perform with them. Our concern is to discover the circumstances under which we can find a smooth p-dimensional subset M of R^n, such that the $f_i(x)$ generate all possible directions that a point can move in M. How does this problem fit in with the problem of reachability? Simple. If such a subset M exists, then starting from a point x_0 in M, the system will be able to move anywhere within M but not out of it by applying the controls $u_i(t)$.

To see this, suppose we start at the point x_0 and follow the trajectory of $dx/dt = f_1(x)$ for t units of time, then $dx/dt = f_2(x)$ for t units, then $dx/dt = -f_1(x)$ for t units, and finally $dx/dt = -f_2(x)$ for t units. This can be done simply by using the constant control inputs $u_i(t) = \pm 1$, as appropriate. Repeated use of the second-order Taylor series expansion around x_0,

$$x(t) = x_0 + t f_i(x_0) + \frac{t^2}{2} \left(\frac{\partial f_i}{\partial x} \right) f_i(x_0) + O(t^3),$$

shows that for t sufficiently small, the resulting point is (see Figure 3.6)

$$x(4t) = x_0 + \frac{t^2}{2}[f_1, f_2]x_0 + O(t^3),$$

where

$$[f_1, f_2] = \frac{\partial f_2}{\partial x} f_1 - \frac{\partial f_1}{\partial x} f_2.$$

The quantity $[f_1, f_2]$ is called the *Lie bracket* of the vector fields f_1 and f_2. From this calculation, we conclude that if $[f_1, f_2]$ is not a linear combination of the vector fields $f_1, f_2, \ldots, f_m$, then $[f_1, f_2]$ represents a "new" direction in which the trajectory can move.

This result provides the mathematical basis for studying the nonlinear reachability problem. Beginning with a set of vector fields $\{f_i(x)\}$, take all linear combinations of these vectors. Then compute the Lie bracket of all elements in this set, taken pairwise. This generates a new

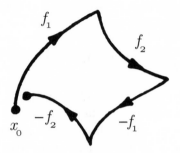

Figure 3.6. Following the vector fields f_1, f_2, $-f_1$, and $-f_2$ from the point x_0.

set of vectors, the objects $[f_i, f_j]$, where $i, j = 1, 2, \ldots, m, i \neq j$. We then take linear combinations of all these elements, throwing them in with the linear combinations obtained before. If this process creates any vectors that are *not* linear combinations of the f_i, we repeat the process, bracketing all the "new" directions with those already present. We continue in this fashion until it no longer produces vectors linearly independent of those already created by this operation of bracketing and formation of linear combinations. The resulting set of vectors is called the *Lie algebra* of vector fields generated by the set $\{f_i\}$, and is denoted $\{f_i\}_{LA}$. Basically, the elements in this Lie algebra define all the directions that one can move from the point x_0. With this result in hand, let's return to the child on the swing.

From the dynamics given earlier,

$$
p(x) = \begin{pmatrix} x_2 \\ -\dfrac{\sin x_1}{x_3} \\ 0 \end{pmatrix}, \qquad g(x) = \begin{pmatrix} 0 \\ -\dfrac{x_2}{x_3} \\ 1 \end{pmatrix}.
$$

These two vectors are clearly linearly independent. Computing the Lie bracket of these two vectors yields

$$
[p, g] = \begin{pmatrix} -\dfrac{2x_2}{x_3} \\ \dfrac{\sin x_1}{x_3^2} \\ 0 \end{pmatrix}.
$$

As long as x_0 is such that its first and second components are not integer multiples of π, this vector is linearly independent of both $p(x)$ and $g(x)$, implying that the Lie algebra generated by $p(x)$ and $g(x)$ fills up the state space. In physical terms, this says that the child can indeed pump the swing to whatever amplitude he or she wishes—provided the swing does not start with zero velocity at the bottom of its arc or begin at the vertical position (presumably with the child falling out of the swing altogether).

This example, simple as it is, shows some of the deep mathematical waters that one quickly gets into when it comes to asking about reachability for general nonlinear processes. The mathematical machinery needed immediately begins to go far beyond elementary linear algebra and matrix theory, calling for a blend of modern differential geometry and abstract algebra—topics usually far beyond the typical comfort zone of even an undergraduate math major. But the situation, while desperate, is not completely hopeless, because if the original problem has some special structure it may still be possible sometimes to remain with the friendly mathematical territory of the linear setting. Let's illustrate this possibility with another example.

Bilinear Control Systems and Reachability

Consider the *bilinear* system

$$\frac{dX}{dt} = \left(F + \sum_{i=1}^{m} N_i u_i(t) \right) X(t), \qquad X(0) = I,$$

where now $X(t)$, F, and the N_i are all $n \times n$ real matrices. Thus, the state space for this system is the set of all such matrices, the so-called *general linear group,* which is denoted $GL(n, R)$. We are interested in characterizing all those matrices that can be "reached" from the identity I using the scalar control inputs $u_i(t)$. Note that this matrix system suffices to check reachability for any vector system of the form

$$\frac{dx}{dt} = \left(F + \sum_{i=1}^{m} N_i u_i(t) \right) x(t), \qquad x(0) = x_0,$$

since $x(t) = X(t)x_0$, for all $x_0 \neq 0$. Note that for this class of systems, our concern is really whether it's possible to reach a state from a nonzero initial state, because if the system starts at the origin, it never leaves—

regardless of the control. This means that the state space of the system is actually the "punctured" space $M = R^n - \{0\}$. In any case, if the matrix system is completely reachable, so is the vector system, and conversely.

As with the example of the child's swing, the answer to the reachability question here hinges on formation of the Lie algebra of matrices generated by the set $\{F, N_1, N_2, \ldots, N_m\}$. This collection of objects is formed in exactly the same way as described earlier for vector fields, using the Lie bracket operation $[F, N] = FN - NF$. Let's denote this algebra by

$$\mathcal{L} = \{F, N_1, N_2, \ldots, N_m\}_{LA}.$$

Then it turns out that the set of reachable states from the identity consists of any $n \times n$ matrix X that can be written as

$$X = e^{A_1} e^{A_2} \cdots e^{A_k}, \quad k = 1, 2, \ldots,$$

where each A_i is in the Lie algebra of matrices $\mathcal{L}$. Here e^A represents the matrix exponential function

$$e^A = I + A + A^2/2! + A^3/3! + \cdots.$$

Let's use this result to test complete reachability for the system:

$$\frac{dx_1}{dt} = x_1(t), \qquad\qquad x_1(0) = x_1^0,$$

$$\frac{dx_2}{dt} = (x_1(t) + x_2(t))(1 + u(t)), \qquad x_2(0) = x_2^0.$$

It is clear by inspection that this system is not completely reachable because the control input has no effect whatsoever on the first component of the state. To verify this fact mathematically, we see that the relevant matrices describing this system are

$$F = \begin{pmatrix} 1 & 0 \\ 1 & 1 \end{pmatrix}, \qquad N = \begin{pmatrix} 0 & 0 \\ 1 & 1 \end{pmatrix}.$$

Computing the Lie brackets to find the Lie algebra $\mathcal{L}$, we obtain

$$[F, N] = \begin{pmatrix} 0 & 0 \\ -1 & 1 \end{pmatrix}, \qquad [F, [F, N]] = \begin{pmatrix} -1 & 0 \\ 0 & 0 \end{pmatrix},$$

$$[N, [F, N]] = \begin{pmatrix} 0 & 0 \\ 0 & 1 \end{pmatrix}, \ldots.$$

It's very easy to see that every element of $\mathcal{L}$ has the form

$$\begin{pmatrix} x & 0 \\ x & x \end{pmatrix},$$

where "x" denotes some possibly nonzero entry. Therefore, the algebra $\mathcal{L}$ can never generate the entire space of 2×2 matrices; hence, our system is not completely reachable.

From this example, we see that if the system under investigation has some special linear structure, then it may be possible to exploit this structure to keep the mathematical machinery within the cozy confines of linear algebra and matrices. But, for the most part, we cannot avoid wheeling out the heavy mathematical artillery previously discussed to deal with the question of reachability and observability for general non-linear processes. Because it would take us way too far afield to enter into a more extensive account of these matters here, we refer the interested reader to the chapter references for the gory details. Now let's leave the mathematical minefield of nonlinearity and return to linear structures and a consideration of the problem of how control inputs affect a system's stability characteristics.

Sharks and Minnows

Suppose we have a population consisting of sharks and a school of minnows, letting $x_1(t)$ represent the minnow population and $x_2(t)$ the number of sharks at time t. Assume that in the absence of sharks, the minnow population grows exponentially, but that the rate of growth of minnows is reduced in direct proportion to the number of sharks present. Furthermore, it is reasonable to suppose that the rate of growth of the shark population is directly proportional to the number of minnows present, but declines exponentially in the absence of minnows, which we are assuming are the only food source for the sharks. Under these conditions, the dynamical equations describing these two populations are

$$\frac{dx_1}{dt} = f_{11}x_1(t) - f_{12}x_2(t),$$
$$\frac{dx_2}{dt} = f_{21}x_1(t) - f_{22}x_2(t),$$

where the f_{ij} are positive constants representing the birth and death rates of the two populations. For notational convenience, let's write these dynamics in the usual vector-matrix form as $dx/dt = Fx$.

It's a straightforward matter to integrate this linear system, obtaining the joint population of sharks and minnows as $x(t) = e^{Ft}x_0$. So what ultimately happens to the population depends on the location of the eigenvalues of F, along with the particular initial population levels x_0. If λ_1 and λ_2 are the eigenvalues of F, we can write the solution of the system as

$$x(t) = a_1 e^{\lambda_1 t} + a_2 e^{\lambda_2 t},$$

where a_1 and a_2 are constant vectors determined by the initial population levels. In particular, this representation shows that both populations die out over time if both eigenvalues are negative, whereas one of them explodes to infinity and the other species dies off if the quantities λ_1 and λ_2 have opposite signs. So the only way there can be a steady state in which both sharks and minnows peaceably coexist is if there is some special relationship between the entries in F and the initial population levels. On the basis of the general arguments to follow, it turns out that a sufficient condition for this to happen is for the initial population ratio $x_1(0)/x_2(0)$ to be greater than the birth ratio f_{11}/f_{22}.

Linear Stability Analysis

Suppose we have the n-dimensional linear system

$$\frac{dx}{dt} = Fx(t), \tag{¶}$$

where F is an $n \times n$ real matrix. The origin is clearly an equilibrium point for the system, so our concern is with the conditions under which it is asymptotically stable. In other words, under what circumstances does a nonzero initial perturbation $x(0) = x_0$ ultimately "wash out"? Or, mathematically, when does $x(t) \to 0$ as $t \to \infty$? The answer to the question is given by the Linear Stability Theorem.

LINEAR STABILITY THEOREM *The equilibrium at the origin for the system (¶) is stable if and only if the eigenvalues of the matrix F all lie in the left half of the complex plane.* ■

For the sake of completeness, let's also state the analogous result for the discrete time linear system

$$x(t + 1) = Fx(t):$$

The origin is an asymptotically stable fixed point for this system if and only if all the eigenvalues of F lie inside the unit circle in the complex plane.

The foregoing results are standard and can be found in any textbook on dynamical systems. But our concern in this chapter is with control systems, which means we are interested in how the behavior of the system's dynamics can be altered by application of control inputs. In particular, suppose the origin is *not* a stable equilibrium. In that case, one or more of the eigenvalues of the matrix F lie in the "bad" half of the complex plane. In such a case, it's certainly reasonable to ask about conditions under which control inputs can make the origin stable.

To get a little insight into what's involved, consider the scalar linear control system

$$\frac{dx}{dt} = fx(t) + gu(t), \qquad x(0) = x_0.$$

It's a simple matter to integrate this system explicitly to obtain

$$x(t) = e^{ft}x_0 + \int_0^t e^{f(t-s)} gu(s)\, ds.$$

Thus, if the constant f is positive, implying that the origin is an unstable equilibrium point of the system, there is generally no control input $u(t)$ that will result in $x(t)$ returning to the origin as the time goes to infinity. So at first glance it doesn't seem as if control can help in stabilizing the uncontrolled process. But not so fast!

In making the above argument, we have essentially chosen the control input $u(t)$ by looking at a clock, then inputting a number $u(t^*)$ into the system dictated solely by the time t^* shown on the clock. This is an example of what control engineers call *open-loop control*. It involves selecting a control input function $u(t)$ over the entire time history of the process (which here is assumed to be $0 \leq t \leq +\infty$), then sitting back and broadcasting the current control value into the system solely on the basis of the current time. This is a time-oriented notion of control. But there is another way.

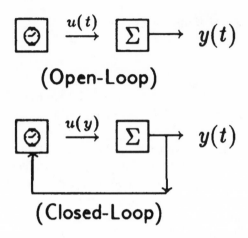

Figure 3.7. Open-loop versus closed-loop (feedback) control.

Suppose, instead of looking at the clock to decide what action to take, we look at what the system itself is doing, thus introducing the concept of *feedback*. This means we measure the state of the system at the current time, and take the control action based upon the outcome of this measurement. So in this case the control input at time t is only implicitly given as a function of time through the state, rather than being given explicitly as before. This is what in engineering parlance is termed *closed-loop (feedback) control*. These two dramatically different notions of control are shown diagrammatically in Figure 3.7.

Mathematically, use of feedback control means that we write the control input as $u(t) = \Phi(x(t))$, where $\Phi(\cdot)$ is the particular feedback function being used. Of special interest to us here is the case of *linear feedback* control, in which we take Φ to be a linear function of the state. This results in the linear feedback control law

$$u(t) = -kx(t),$$

where k is a real number. The minus sign has been inserted here to conform with standard conventions in control engineering. Using this linear feedback control, the dynamical equations governing the behavior of the controlled system are

$$\frac{dx}{dt} = (f - gk)x, \qquad x(0) = x_0.$$

Now look what happens if f is positive. Simply by choosing the feedback constant to be $k = (f-a)/g$, we can arrange for the controlled system's dynamics to be

$$\frac{dx}{dt} = ax,$$

where a can be taken to be *any* real number that we wish. In particular, taking a to be a negative real number converts the unstable equilibrium at the origin into one that's now asymptotically stable.

Examining what makes the above stabilization by feedback magic work, we see that there is just a single condition: the control coefficient g must be nonzero. Physically, this means merely that the control input can actually act so as to influence the system state—a condition that we discussed earlier as being the sine qua non of control theory. In this scalar case, the condition $g \neq 0$ boils down to the system being what we earlier termed completely reachable. From here it's but a small speculation to wonder if in the more general case of an n-dimensional system with m control inputs, we could arbitrarily alter the position of the eigenvalues of the state matrix by application of linear feedback if the system is completely reachable. Unfortunately, the situation is not quite *that* good—but almost. Here's the story.

We begin with the n-dimensional linear control system with m inputs

$$\frac{dx}{dt} = Fx(t) + Gu(t).$$

Let's agree to use the linear feedback control law $u(t) = -Kx(t)$, where K is a constant $m \times n$ real matrix. If we let $\lambda_1, \lambda_2, \ldots, \lambda_n$ denote the eigenvalues of the system dynamics matrix F, we want to know how these quantities can be changed by the choice of K. The following result answers this question.

POLE-SHIFTING THEOREM *Let* $\Gamma = \{\gamma_1, \gamma_2, \ldots, \gamma_n\}$ *be a set of complex numbers such that if* γ *is in* Γ, *then its complex conjugate* $\bar{\gamma}$ *is also in* Γ. *Furthermore, assume the system is completely reachable. Then there exists a unique feedback matrix* K *such that the eigenvalues of the closed-loop dynamics matrix* F − GK *are exactly the set* Γ. ∎

There are several points worthy of note about this basic result.

1. The theorem says that under the very weak constraint of completely reachability, the closed-loop dynamics can, for all intents and purposes, be prescribed arbitrarily. This fact is of the utmost significance from a system design point of view because it allows the designer to set up the dynamics more or less at will, safe in the knowledge that any type of stability behavior can be achieved after the fact through an appropriately chosen feedback control.

2. The condition on the elements of the set Γ means that Γ is what is called a *symmetric set* of complex numbers. This restriction can be relaxed, provided we are willing to entertain *complex,* rather than real, feedback matrices K. Mathematically, this is no problem; but practically speaking, it's a bit of a nuisance trying to construct devices to actually implement a complex-valued control input.

3. The term "pole-shifting" arises from the standard electrical engineering representation of a linear system as a rational function (a quotient of two polynomials) instead of as a dynamical system as we have done here. In the rational function representation, the eigenvalues of F coincide with the zeros of the denominator polynomial, which in the jargon of complex function theory are the "poles" of the rational function. So what the Pole-Shifting Theorem says is that we can relocate the poles of the uncontrolled system to wherever we want them to be in the complex plane simply by using an adroitly chosen linear feedback control law.

Now let's see the Pole-Shifting Theorem in action.

Suppose we have a single-input system with dynamics specified by the matrices

$$F = \begin{pmatrix} -1 & 2 & 1 \\ -2 & 3 & -1 \\ 1 & -1 & 0 \end{pmatrix}, \qquad G = \begin{pmatrix} 0 \\ 1 \\ 0 \end{pmatrix}.$$

A short calculation shows that the eigenvalues of F are $-1, 1$, and 2. This means that the origin is unstable for this system because two of the eigenvalues of F are in the right half of the complex plane.

Let's now employ the linear feedback control $k = (k_1, k_2, k_3)$, and try to choose the numbers k_i so that the closed-loop system matrix $F - gk$ has eigenvalues $-1, -2$, and -3. Can we find such a choice?

First, we check that the pair (F, G) is completely reachable. The reachability matrix

$$C = \begin{pmatrix} 0 & 2 & 3 \\ 1 & 3 & 6 \\ 0 & -1 & -1 \end{pmatrix}$$

has three linearly independent columns, which means that the system is indeed completely reachable. Moreover, the set $\Gamma = \{-1, -2, -3\}$ is symmetric by default because it contains only real numbers. Thus, the conditions for the Pole-Shifting Theorem are satisfied, ensuring the feedback law we seek exists. It is left to the reader to verify that the choice $k_1 = -8, k_2 = 12$, and $k_3 = 12$ does the trick. For a more detailed discussion of exactly how to compute these quantities, the reader is invited to consult the material cited in the chapter references.

Up to now, our attention has been focused on what is *feasible* by way of influencing a system via control inputs. We have studiously avoided any discussion of what inputs might be "optimal" with respect to any kind of criterion. For instance, there may be many control inputs that will transfer a system from the origin to some desired end state. So once we have demonstrated the existence of at least one such needle in the haystack, it makes sense to speak about trying to find the "best" needle, if there are more than one. The remainder of the chapter is devoted to a consideration of such matters.

What's Best?

In 1696, mathematician Johann Bernoulli posed the problem illustrated in Figure 3.8. It involves a bead sliding in a plane along a frictionless wire having fixed endpoints P and Q. Bernoulli asked for the shape of the wire that causes the bead to move from P to Q under the influence of gravity in the shortest possible time. This is the so-called *brachistochrone problem*, whose solution is the cycloid shown in the figure.

The brachistochrone problem was the forerunner of what is now known as the *calculus of variations*, by which we seek curves that cause some criterion function to attain either its minimal or maximal value. These kinds of problems, in turn, form the basis for the theory of *optimal control*.

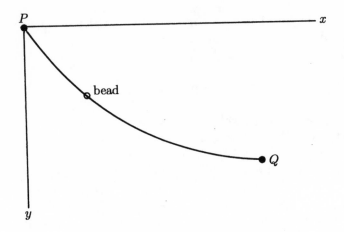

Figure 3.8. The brachistochrone problem.

We begin by considering what is termed the *simplest problem in the calculus of variations*. It involves finding a curve $x(t)$ over an interval $0 \leq t \leq T$ that minimizes the integral

$$J(x) = \int_0^T \phi(x, dx/ds, s) \, ds,$$

subject to the condition that $x(0) = c$. It's an easy matter to see that this problem is equivalent to finding a control function $u(t)$ that minimizes

$$J(u) = \int_0^T \phi(x, u, s) \, ds,$$

subject to the differential equation side constraint

$$\frac{dx}{dt} = u, \qquad x(0) = c.$$

So, we see that the simplest problem in the calculus of variations corresponds to a very special type of optimal control problem.

To find the optimizing curve $x^*(t)$ (or, equivalently, the minimizing control function $u^*(t)$), we proceed formally, assuming that such a minimizing function exists. Then, if $z(t)$ is any "nearby" function, we must have

$$J(z) \geq J(x^*).$$

Since $z(t)$ is a nearby function, we write

$$z(t) = x^*(t) + \epsilon q(t),$$

where $q(t)$ is an as yet unspecified function and ϵ is a small parameter. The inequality now becomes

$$J(x^* + \epsilon q) \geq J(x^*),$$

which holds for *all* ϵ and *all* functions $q(t)$ nearby $x^*(t)$.

Expanding the left-hand side of the above inequality in powers of ϵ, we obtain

$$J(x^*) + \epsilon \left[\int_0^T (\phi_x q(s) + \phi_{x'} q'(s)) \, ds \right] + O(\epsilon^2) \geq J(x^*),$$

which holds for all sufficiently small ϵ. Here we have used the notation

$$\phi_x \doteq \partial\phi/\partial x, \qquad \phi_{x'} \doteq \partial\phi/\partial x'.$$

For this inequality to be valid for both positive and negative values of ϵ, we must have

$$\int_0^T (\phi_x q(s) + \phi_{x'} q'(s)) \, ds = 0.$$

But since $q(t)$ was arbitrary, this relation must hold for all differentiable functions $q(t)$.

By performing integration by parts on the second term in the above integrand, we get

$$\int_0^T q(s) \left[\phi_x - \frac{d}{ds}\phi_{x'} \right] ds + q(s)\phi_{x'} \Big|_0^T = 0.$$

If we agree to keep the endpoints of the optimizing curve x^* fixed, then $q(0) = q(T) = 0$, and we have

$$\int_0^T q(s) \left[\phi_x - \frac{d}{ds}\phi_{x'} \right] ds = 0,$$

for all admissible perturbation functions $q(t)$. An appeal to the Fundamental Lemma of the Calculus of Variations now allows us to conclude that

$$\phi_x - q(s)\frac{d}{dt}\phi_{x'} = 0.$$

133

This equation is termed the *Euler-Lagrange equation,* which, together with the boundary conditions $x^*(0) = c, dx^*(T)/dt = 0$, constitutes a necessary condition that any candidate for the optimizing curve must satisfy.

To see how this formalism works, suppose we take

$$\phi(x, u, t) = \tfrac{1}{2}(x^2 + u^2).$$

Then

$$\phi_x = x \quad \text{and} \quad \phi_{x'} = \phi_u = u \ \left(= \frac{dx}{dt} \right),$$

which leads to the Euler-Lagrange system

$$x - \frac{d}{dt}\left(\frac{dx}{dt} \right) = 0,$$

or

$$\frac{d^2x}{dt^2} - x = 0,$$

with the boundary conditions $x(0) = c, \ dx(T)/dt = 0$. This is the two-point boundary-value problem that must be satisfied by the minimizing curve $x^*(t)$. In this case the equation can be solved explicitly, leading to the minimizing curve

$$x^*(t) = c\frac{\cosh(t - T)}{\cosh T},$$

along with the corresponding optimal control

$$u^*(t) = dx^*/dt = c\frac{\sinh(t - T)}{\cosh T}.$$

It is important to note that the Euler-Lagrange equation is only a *necessary* condition for a curve to be an extremal—a first filter, so to speak—and very far from being sufficient. To illustrate this point, consider the problem of minimizing

$$J(x) = \int_0^{4\pi} [(dx/dt)^2 - x^2]\, dt, \qquad x(0) = 0.$$

The Euler-Lagrange equation is readily seen to be

$$\frac{d^2x}{dt^2} + x = 0,$$

with the boundary conditions

$$x(0) = 0, \qquad \frac{dx}{dt}\bigg|_{t=4\pi} = 0.$$

This equation possesses the unique solution $x(t) = 0$. However, the minimum value of $J(x)$ is *not* zero.

To see this, consider the trial function $x(t) = k \sin t/2$. After an integration by parts, we have

$$J\left(k\sin\frac{t}{2}\right) = -\frac{3k^2}{4}\int_0^{4\pi} \sin^2\left(\frac{t}{2}\right) dt.$$

Thus, as k increases without bound, the integral assumes arbitrarily large negative values; hence, there is no minimum, despite the fact that the Euler-Lagrange equation has a unique solution.

The problem with the foregoing example is that the interval length $[0, 4\pi]$ is too long. It turns out to be an easy exercise to show that if we consider the same problem over an interval $[0, T]$, where T is sufficiently small, then the integral has the absolute minimum zero over the class of square-integrable functions $x(t)$. The point to note here is that satisfaction of the Euler-Lagrange equation does not suffice to pin down the minimizer; for this, other filters such as the Weierstrass condition, the Legendre condition, the Transversality condition and so on are needed. The reader can find accounts of these tests in the chapter references.

The arguments given above have all been for the special control system

$$\frac{dx}{dt} = u(t).$$

In the early 1960s, L. S. Pontryagin and his coworkers at the Steklov Mathematical Institute in Moscow established a wide-sweeping generalization of these classical ideas to the case of general control processes. Their results, now collectively termed the *Minimum Principle,* enable us to extend the above arguments to a very broad class of optimal control problems. Let's have a look at what the Moscow school accomplished.

The Pontryagin Minimum Principle

The prototypical optimal control process we wish to consider here is that of finding a control function $u(t)$, $0 \leq t \leq T$, such that the integral

$$J(u) = \int_0^T g(x, u) \, dt$$

is minimized, where the system state $x(t)$ is connected to the control input $u(t)$ via the differential equation

$$\frac{dx}{dt} = f(x, u), \qquad x(0) = x_0.$$

There are many variations upon the basic theme of this problem, including those involving assumptions about the class of admissible control inputs, constraints on the state, the degree of smoothness in the functions f and g, whether the interval length T is finite or infinite, and other types of criteria such as minmax and so forth. But for the sake of exposition and to avoid annoying and distracting technical caveats, we assume here that the functions f and g, as well as the space of admissible inputs and the constraint space on the state, possess whatever mathematical properties we need in order to have our formal manipulations make sense.

To find the function $u^*(t)$ minimizing the integral J under the differential equation side constraint, the Pontryagin approach involves introducing what's called a *Lagrange multiplier* function $p(t)$. This maneuver enables us to move the side constraint (the system dynamics) into the cost functional J, leading to the Hamiltonian function for the system as

$$H(x, u, p, t) = g(x, u) + (p(t), f(x, u)),$$

where again $(\cdot, \cdot)$ is the usual vector inner product. The Pontryagin group then proved the following surprising fact:

PONTRYAGIN MINIMUM PRINCIPLE *The optimal control* u*(t) *is that function* u *minimizing the Hamiltonian function* H *pointwise in* t. ■

As before, we let the control input $u(t)$ take its values in R^m, with the state $x(t)$ being a point in R^n. Using the *co-state* vector $p(t)$, we

136

form the Hamiltonian function H as before. So H is a scalar-valued function of $x, u,$ and p. By the Minimum Principle, the minimizing control function $u^*(t)$ is exactly that admissible input function u that minimizes H.

Of course, the function H contains the unknown co-state function p. So to find the minimizing control, we must develop a relation linking the control function $u(t)$ to the co-state function $p(t)$. Following an approach very similar to that employed to obtain the classical Euler-Lagrange equation, the Pontryagin group showed that the functions $x, u,$ and p are related to each other through the following two-point boundary-value problem:

$$\frac{dx}{dt} = \frac{\partial H(x, u, p, t)}{\partial p}, \qquad x(0) = x_0,$$

$$-\frac{dp}{dt} = \frac{\partial H(x, u, p, t)}{\partial x}, \qquad p(T) = 0, \qquad 0 \le t \le T.$$

Assuming there are no constraints imposed on the allowable variations in the control u, a necessary condition for the optimizing input function u^* is that it satisfy the equation

$$\frac{\partial H(x, u, p, t)}{\partial u} = 0,$$

where the boundary-value problem given above is used to express both x and p as functions of u.

We have assumed that the final time T is fixed but that there are no conditions imposed on the final state $x(T)$. This leads to the boundary condition given previously for the co-state function $p(t)$ at $t = T$. If the final time is not fixed, determination of this boundary condition is a bit trickier, leading to what are termed *transversality conditions*. Details on these matters are found in the texts cited in the chapter references. Now let's see the Minimum Principle in action by returning to the problem of the Viking lander sketched at the beginning of this chapter.

Minimal-Time Soft Landing

Suppose the Viking lander is in its terminal phase, trying to make a soft landing on the surface of Mars in minimal time. The diagram in

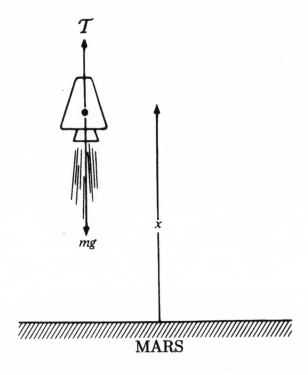

Figure 3.9. Soft landing for the Viking lander.

Figure 3.9 shows the overall situation, where m denotes the mass of the lander, g the acceleration due to gravity, and T the thrust.

To model this situation, we make the following simplifying assumptions:

1. Aerodynamic forces, as well as gravitational forces of bodies other than Mars, are negligible.
2. Lateral motion is ignored; consequently, the descent trajectory is vertical and the thrust T is along the line of the lander's trajectory.
3. Due to the nearness of the lander to the Martian surface, the acceleration due to gravity is constant.
4. The relative velocity of the exhaust gases with respect to the lander is constant.

Under these conditions, the rate of change of the lander's mass is constrained to be $-M \le dm/dt \le 0$, where M is a maximal rate at which the mass can be shed by burning fuel to produce thrust. The dynamic equations for the lander are given by Newton's Second Law as

$$m\frac{d^2x}{dt^2} = -gm(t) + T(t)$$

$$= -gm(t) - k\frac{dm}{dt},$$

since the controlling thrust is just proportional to the rate at which the lander's mass is reduced. Here the constant k represents the relative exhaust velocity of gases with respect to the lander.

If we define the system states to be $x_1 = x$, the lander's position; $x_2 = dx/dt$, the velocity; and $x_3 = m$, the mass; and if the control is given by $u = dm/dt$, the rate at which mass is burned, then the control system can be written as

$$\frac{dx_1}{dt} = x_2(t), \qquad\qquad x_1(0) = x_0,$$

$$\frac{dx_2}{dt} = -g - ku(t)/x_3(t), \qquad x_2(T) = 0,$$

$$\frac{dx_3}{dt} = u(t), \qquad\qquad x_3(T) = 0.$$

The cost criterion is simply that the time of burn be as short as possible. This translates into choosing the burn so as to minimize

$$\int_0^T 1\,ds.$$

With this criterion of optimality, the Hamiltonian for the system is

$$H(x, u, p, t) = 1 + p_1(t)x_2(t) - gp_2(t) - kp_2(t)u(t)/x_3(t) + p_3(t)u(t).$$

The Pontryagin Minimum Principle now asserts that the optimal control $u^*(t)$ carrying the lander to a soft landing in minimal time must satisfy

$$H(x^*(t), u^*(t), p^*(t), t) \leq H(x^*(t), u(t), p^*(t), t)$$

for all admissible thrust patterns $u(t)$ and all t.

It is clear from the fact that the control $u(t)$ appears linearly in the Hamiltonian that the structure of the minimal-time controller must be

$$u^*(t) = \begin{cases} 0, & \text{for } p_3^*(t) - kp_2^*(t)/x_3^*(t) < 0, \\ -M, & \text{for } p_3^*(t) - kp_2^*(t)/x_3^*(t) > 0, \\ \text{undetermined}, & \text{for } p_3^*(t) - kp_2^*(t)/x_3^*(t) = 0. \end{cases}$$

What this says is that the fastest way to get the lander onto the Martian surface is to use what's called a *bang-bang* control policy; that is, either burn at the maximal rate of thrust or don't burn at all. Of course, to know exactly when to switch between these two extremes, we have to solve a nonlinear two-point boundary-value problem to obtain the optimal state $x^*(t)$ and co-state $p^*(t)$.

The Pontryagin Minimum Principle leads to the solution of the optimal control problem in the form of what we earlier termed an "open-loop" control law. That is, a control function given explicitly in terms of time: look at the clock, and then apply the control called for at that instant of time. But in examining the stability of a control system, we saw that there was much to be gained by having the control law given in "closed-loop," or "feedback," form. In that case, we could actually change the dynamics of the system through the control input. So it's natural to wonder about what type of procedure might enable us to obtain the optimal control input in feedback form. The answer comes in the form of what is called *dynamic programming*.

Feedback Control and Dynamic Programming

The key ingredient in the dynamic programming approach to optimal control is Bellman's Principle of Optimality. Assume that $u^*(t)$ is the optimal control function and that $x^*(t)$ is the associated optimal state trajectory. Let $v = u^*(0)$ denote the optimal action to be taken at the initial moment, with the initial state being $x(0) = c = x^*(0)$. Then the Principle of Optimality states that the part of the optimal trajectory starting at time $t = \Delta$ from the state $c + f(c, v, 0)\Delta$ is also the optimal trajectory for a problem that begins not at time $t = 0$ in the state c, but at time $t = \Delta$ in the state $c + f(c, v, 0)\Delta$. In other words, any part of an optimal state trajectory is also an optimal trajectory. The general idea is illustrated in Figure 3.10.

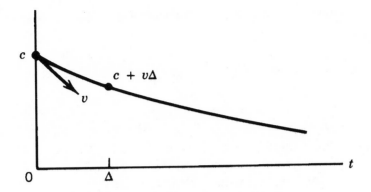

Figure 3.10. The Principle of Optimality.

Let's use the Principle of Optimality to develop an equation for the control that minimizes

$$J = \int_t^T \phi(y, u)\, dt + \psi(x(T)),$$

where

$$\frac{dx}{dt} = f(x, u), \qquad x(t) = c,$$

and the observed output is

$$y(t) = h(x(t)).$$

Note that this formulation of the problem agrees with that given earlier if we set $t = 0$ and $c = x_0$. So what we are doing here is imbedding the original problem within a *family* of problems parameterized by the initial time t and the initial state c. The minimal value of J, as well as the associated optimal control law, is determined by two quantities: the initial state c and the length of time of the process, which is characterized by $T - t$ where T is assumed to be fixed.

Begin by introducing the function

$I(c, t) =$ the value of J obtained when an optimal control is used
for a process of duration $T - t$ that starts in state c.

Then regardless of what initial control action $v = u(0)$ is applied, the Principle of Optimality, together with the definition of the function

$I(c, t)$, shows that we must have the inequality

$$I(c, t) \le \{\phi(h(c), v)\Delta + I(c + f(c, v)\Delta, t + \Delta)\} + o(\Delta). \qquad (\S)$$

Here we have made use of the additivity property of integrals

$$\int_t^T = \int_t^{t+\Delta} + \int_{t+\Delta}^T,$$

as well as the Mean-Value Theorem for integrals. We clearly want to make the right side of ($\S$) as small as possible. So we choose v to minimize it, leading to the recurrence relation for I as

$$I(c, t) = \min_v \{\phi(h(c), v)\Delta + I(c + f(c, v)\Delta, t + \Delta)\} + o(\Delta). \quad (\S\S)$$

For a process of length $T - t = 0$, we have the trivial starting condition

$$I(c, T) = \psi(c).$$

The relation ($\S\S$) links the solution to a problem starting at time t with one starting at time $t + \Delta$. So this recurrence relation, together with the initial condition $I(c, T) = \psi(c)$, enables us to calculate the *optimal-value function* $I(c, t)$, as well as the optimal decision function $v(c, t)$, for each possible initial state c and every interval length $T - t$. By this approach, we solve not only the original problem involving some *fixed* values of c and t (such as $c = x_0, t = 0$), but at the same time solve a host of additional problems having different initial states and various interval lengths.

If we pass to the limit letting $\Delta \to 0$ in Eq. ($\S\S$), we obtain the famous Bellman-Hamilton-Jacobi equation for the function I. It is

$$-\frac{\partial I}{\partial t} = \min_v \left\{\phi(h(c), v) + \left(f(c, v), \frac{\partial I}{\partial c}\right)\right\}, \qquad t < T,$$

$$I(c, T) = \psi(c).$$

With these ideas about how dynamic programming works in principle, let's look at an important example that will serve as our touchstone for the rest of the chapter.

Linear Dynamics, Quadratic Costs

In the last section we showed how to solve a quadratic optimization problem with scalar linear dynamics using the Minimum Principle. Now we return to the same problem, but in its full n-dimensional form, showing how the solution looks when we use the Bellman-Hamilton-Jacobi equation and dynamic programming. Recall that the functions defining the cost criterion and the dynamics of this linear-quadratic (LQ) problem are

$$\phi(y, u) = \tfrac{1}{2}[(y, Qy) + (u, Ru)], \qquad \psi(x) = (x, Mx),$$
$$f(x, u) = Fx + Gu, \qquad h(x) = Hx,$$

where x is in R^n, u is in R^m, and y is in R^p. To ensure that there exists a unique minimizing control, we assume that Q and M are $n \times n$ positive-semidefinite matrices and R is an $m \times m$ positive-definite matrix.

The Bellman-Hamilton-Jacobi equation for in this case becomes

$$-\frac{\partial I}{\partial t} = \min_v \left\{ \tfrac{1}{2}[(Hc, QHc) + (v, Rv)] + \left(Fc + Gv, \frac{\partial I}{\partial c} \right) \right\},$$
$$t < T,$$

$$I(c, T) = (c, Mc).$$

It's now a textbook exercise to verify that the function $I(c, t)$ is a pure quadratic form in c. That is, there exists an $n \times n$ matrix function $P(t)$ such that

$$I(c, t) = (c, P(t)c).$$

Some tedious, although straightforward, algebra now yields

$$v_{\min}(c, t) = u^*(t) = -R^{-1}G'P(t)c,$$

where the function $P(t)$ satisfies the matrix Riccati equation

$$-\frac{dP}{dt} = H'QH + PF + F'P - PGR^{-1}G'P, \qquad P(T) = M.$$

The optimal state trajectory is the solution of the *closed-loop* dynamics

$$\frac{dx^*}{dt} = [F - GR^{-1}G'P(t)]x^*(t), \qquad x^*(t) = c.$$

The most important point to note about this example is that the optimal control is expressed as a function of the system state c, rather than as a function of the current time t. This means that we must *measure* what the system is actually doing and apply the control that is appropriate for that state, rather than just look at a clock and determine the control as a function of the time alone. Basically, the state measurement is "fed back" from the system output to its input, giving rise to the concept of *feedback*, or *closed-loop*, control discussed earlier.

We'll come back to the LQ problem with a vengeance in just a moment. But first, let's briefly compare the Pontryagin and the Bellman approaches to the solution of optimal control problems.

The Minimum Principle versus Dynamic Programming

The calculus of variations and dynamic programming are complementary theories. The former regards the extremal curve as a locus of points, and attempts to determine the curve as the solution of a differential equation, the Euler-Lagrange equation. The theory of dynamic programming looks at the extremal curve as an envelope of tangents and seeks the optimal direction to move at each point on the extremal. The basic duality of Euclidean geometry states that a curve can equally be thought of as a locus of points of an envelope of tangents. As a result, the two theories that we have presented—the calculus of variations and dynamic programming—are dual to each other in exactly this geometrical sense. These two complementary points of view are illustrated in Figure 3.11.

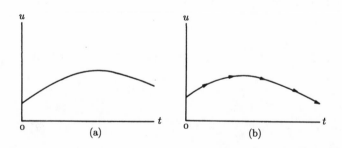

Figure 3.11. (a) A locus of points. (b) An envelope of tangents.

It should also be noted that the basic equation for the optimal-value function I is a partial differential equation rather than the ordinary differential equation obtained using the Minimum Principle. However, instead of being a boundary-value problem, the dynamic programming equation for I is an *initial-value* problem, which is generally easier to deal with computationally than a boundary-value problem. So we can trade off ordinary differential equations and boundary-value problems for partial differential equations and initial-value problems. In general, which method we use depends upon a host of circumstances surrounding the specific form of the problem, the computing capability at our disposal, and a lot of other concerns that space constraints prevent us from entering into here. The interested reader is invited to consult the many excellent references on this circle of questions cited in the chapter references.

We have now seen duality enter our deliberations twice: the duality of completely reachable and completely observable systems, as well as the duality of the calculus of variations and dynamic programming. As we have just seen, the LQ control problem can be solved by both the Minimum Principle and dynamic programming. As a control problem, it's basically involved with what can be done by way of making the system behave in certain ways by exerting control, that is, reachability. But reachable systems are the mathematical duals of observable ones, so it's not out of place to ask about the mathematical dual to the LQ control problem. Intuition suggests that whatever this dual problem turns out to be, it will involve the notion of observation in a basic way. And so it is. In fact, it involves observation in such a basic way that the LQ control problem dual ends up being one of the two or three most useful pieces of mathematics developed in this century. Let's see why.

Getting the Best of It

Suppose we have a dynamical process governed by the equation

$$\frac{dx}{dt} = f(x, t) + \text{dynamical error}, \qquad x(0) = c,$$

giving rise to the observed output

$$y(t) = h(x, t) + \text{observational error}.$$

Here, for simplicity, we assume that f and h are scalar-valued functions, and that the initial state c is known exactly. The task is to estimate the state of the process over some finite period of time, $0 \leq t \leq T$.

Suppose we agree to let the error term in the dynamics be represented by a function $u(t)$, and select this function so as to produce the optimal estimate of the state. Thus, the function $u(t)$ accounts for our uncertainty about the true dynamics and for whatever stochastic effects may be influencing the state. It is clear that, on the one hand, we want to choose $u(t)$ so that the state follows a trajectory reasonably close to that dictated by what we think the state dynamics actually are (given by the vector field f). On the other hand, we don't want our estimate of the state to depart too wildly from what has actually been observed (the function $y(t)$). So we need to choose the control function to optimally trade off these two costs.

Measuring the costs by a least-squares criterion, we are led to the control problem:

$$\min_{z} \int_0^T \left[k_1(t)\{y(t) - h(x(t))\}^2 \right.$$
$$\left. + k_2(t)\{dz/dt - f(z(t), t)\}^2 \right] dt + k_3 x^2(T),$$

where $k_1(t)$ and $k_2(t)$ are nonnegative weighting functions representing our relative confidence in the dynamics f and the observations y. So if we feel more confident in the dynamics than in the observations, we make $k_1(t)$ large and $k_2(t)$ small, and conversely. Finally, the nonnegative number k_3 measures how important it is that the system end up near zero.

We can recast this formulation into more familiar terms by recalling that the control function $u(t)$ was introduced to account for dynamical error. That is, we have assumed that the *true* dynamics are

$$\frac{dx}{dt} = f(x, t) + u(t).$$

Thus, the optimal control problem becomes

$$\min_{u} \int_0^T \left[k_1(t)\{y(t) - h(x(t))\}^2 + k_2(t)u^2(t) \right] dt + k_3 x^2(T),$$

subject to the constraint

$$\dot{x} = f(x, t) + u, \qquad x(0) = c.$$

Once the weighting functions $k_1(t)$ and $k_2(t)$ and the constant k_3 have been given, we can use either the Minimum Principle or dynamic programming to solve this problem in the manner outlined above. We'll return to this purely *deterministic* control problem shortly, after looking at a seemingly unrelated *stochastic* estimation problem.

State Estimation

Consider an n-dimensional system whose output is observed at times $k = 0, 1, 2, \ldots$. Each observation is corrupted by noise, however, so what's actually seen at time k is

$$y_k = Hx_k + v_k,$$

where H is a $p \times m$ output matrix and v_k is the noise in the observation at time k. What we seek is the "best" estimate for the state x_k based on the observed values $y_0, y_1, \ldots, y_k$ and whatever knowledge is available on the statistical properties of the noise.

As just stated, the problem is unsolvable. The missing element is some link between the current state x_k with the previous state x_{k-1}. In Kalman filtering, this linkage is assumed to be linear, with an additional "noise" term introduced to account for uncertainties in the system dynamics. Thus,

$$x_{k+1} = Fx_k + Gu_k,$$

where F and G are $n \times n$ and $n \times m$ matrices, respectively. Moreover, the noise u_k is assumed to have known (or postulated) statistical properties. In fact, the standard assumption in Kalman filtering is that both the observation noise and the noise in the dynamics are "white" Gaussian processes with zero mean. This means that everything you need to know (or can know) about the noises v_k and u_k is contained in their covariance matrices.

What Rudolf Kalman showed in 1960 was that the best estimate of the state $\hat{x}_k$, along with the covariance matrix for the error in this best state estimate, could be obtained recursively from the previous best state estimate and its error covariance matrix. In other words, the Kalman filter uses each new observation to update a probability distribution for the state of the system, never having to refer back to any earlier observation. This is analogous to updating a running average A_n of numbers

$a_1, a_2, \ldots$, by the recursive relationship

$$A_n = \left(\frac{n-1}{n}\right) A_{n-1} + \left(\frac{1}{n}\right) a_n,$$

instead of starting over again each time with the formula $A_n = (a_1 + a_2 + \ldots + a_n)/n$.

The significance of this recursion relation for the best state estimate is that the Kalman filter does no more work for the billionth estimate than it does for the first. Furthermore, the error covariance computations do not depend on the observed values y_k, but involve only the covariance matrices of the two noise processes v_k and u_k. Thus, this part of the filter can be precomputed, leading to an algorithm that is ideally suited for real-time applications for which the data arrive sequentially.

The Kalman filter is used in just about every inertial navigation system in existence. For example, Hexad, the gyroscopic system on the Boeing 777 aircraft, uses a Kalman filter to estimate the errors in each of the six gyros with respect to the others. Because only three gyros suffice for navigation, a single malfunctioning component can be easily identified by the Kalman filter and safely ignored.

Important as inertial navigation is, it is not the only area to which the Kalman filter makes a contribution. In fact, anything that moves, if it's automated, is a candidate for a Kalman filter. One possible use arises in what is called "active noise control." This involves removing unwanted sounds from the environment coming from things such as an airplane engine or your neighbor's stereo. The idea is to create zones of quiet by broadcasting "antinoise" to cancel the unwanted noise. A Kalman filter is needed to estimate the coefficients used to construct the antinoise.

Historically, the Kalman filter was used, for example, in the on-board computer that guided the descent of the *Apollo 11* lunar module to the Moon. That computer was communicating with a system of four radar stations on Earth that were monitoring the module's position. It was important, of course, that the estimates for all sources be good. The Earth-based estimates were used to adjust the on-board system; if there had been too much disagreement, then the mission would have been scrubbed. Since that time, the Kalman filter has found its way into data streams everywhere, from seismology to econometrics. So with this advertising hype for the penetration of the Kalman filter into just about every nook and cranny of science, let's finally look at the details.

The Kalman Filter

To make as direct a comparison as possible with the LQ control problem discussed above, we'll assume that we are dealing with observations arising in continuous time. Specifically, we have a message generated by the model

$$\frac{dx}{dt} = Fx(t) + Gu(t), \qquad x(0) = x_0.$$

Here the message is the state $x(t)$, which is corrupted by the random input $u(t)$. We'll say more about the nature of this noise in a moment. The actual observed signal is

$$z(t) = Hx(t) + v(t),$$

where $v(t)$ is another random process representing the noise in the observations. We make the following assumptions about the noise processes $u(t)$ and $v(t)$:

1. The noise processes u and v are Gaussian, with mean zero and covariance matrices Q and R, respectively. In mathematical parlance, this means that

$$E[u(t)] = E[v(t)] = 0,$$
$$E[u(t)u'(\tau)] = Q\delta(t - \tau),$$
$$E[v(t)v'(\tau)] = R\delta(t - \tau),$$

   for all t, τ, where $\delta(\cdot)$ is the Dirac delta function, Q is a positive semidefinite matrix, and R is a positive definite matrix. These latter conditions on Q and R mean that some components of the state may be noise-free, but that *all* the observations are corrupted by some noise.

2. The processes $u(t)$ and $v(t)$ are uncorrelated; in other words, $E[u(t)v'(\tau)] = 0$ for all t, τ.

3. The initial condition x_0 is a Gaussian random variable with mean zero and covariance matrix M; that is, $E[x_0] = 0$, while $E[x_0x_0'] = M$. Moreover, x_0 is independent of both of the noise processes u and v; that is, $E[x_0v'(t)] = E[x_0u'(t)] = 0$ for all $t \geq 0$.

With these preliminaries, the optimal linear filtering problem becomes: Given the measured values $z(t)$ in the interval $0 \leq t \leq T$, find an estimate $\hat{x}(t)$ of the state of the form

$$\hat{x}(t) = \int_0^t A(t, s) z(s) \, ds,$$

where A is an $n \times p$ matrix function termed the *impulse response* for the filter. The estimate $\hat{x}(t)$ is to be made so as to minimize the quadratic cost

$$E[(\hat{x}(t) - x(t))(\hat{x}(t) - x(t))'].$$

In other words, we want the expected square of the deviation between the estimate $\hat{x}(t)$ and the true state $x(t)$ to be as small as possible. (*Note:* Demanding that the estimate $\hat{x}(t)$ be given in the integral form means that we are asking for the best *linear* estimator of the state. Under the hypotheses stated here about the message model, the observations, and the noise processes, it can be shown that there is no better estimator than the linear one expressed above. But if any of these conditions are relaxed, then a nonlinear estimate may do better. But these are matters beyond the scope of our discussion here.)

The solution to this estimation problem is not given by an explicit formula for the impulse response matrix $A(t, s)$, but rather by a recursive scheme for actually calculating the optimal estimate $\hat{x}(t)$. This scheme is the celebrated

---

KALMAN FILTER THEOREM   *The optimal estimate of the state* $\hat{x}(t)$
satisfies the dynamical equation

$$\frac{d\hat{x}}{dt} = F\hat{x}(t) + K(t)[z(t) - H\hat{x}(t)],$$
$$\hat{x}(0) = 0,$$

where

$$K(t) = P(t) H R^{-1}$$

is determined by the function P(t) *satisfying the matrix Riccati equation*

$$\frac{dP}{dt} = GQG' + FP(t) + P(t)F' - P(t)H'R^{-1}HP(t),$$
$$P(0) = M.$$

150

Moreover, the optimal error in the estimate, x̃(t) = x(t) − x̂(t), satisfies the equation

$$\frac{d\tilde{x}}{dt} = F\tilde{x}(t) + Gu(t) - K(t)[H\tilde{x}(t) + v(t)],$$
$$\tilde{x}(0) = x_0. \quad \blacksquare$$

So we see that the Kalman filter is a feedback system. It is obtained by taking a copy of the message process (without the system noise $u(t)$), forming the error between the actual observation $z(t)$ and what would be obtained by using the best estimate $\hat{x}$ (i.e., the quantity $\tilde{z}(t) = z(t) - \hat{z}(t) = H[x(t) - \hat{x}(t)]$) and feeding this error back into the system, weighted by the so-called *Kalman gain matrix* $K(t)$. As noted above, the importance of this result cannot be overemphasized because it enables us to calculate the optimal state estimate as we go along.

To see how this setup works in practice, let's return to the case considered at the beginning of the chapter of the spacecraft moving along a line. Suppose the spacecraft leaves the origin at time $t = 0$ with a fixed, but unknown, velocity. Let's assume the unknown velocity is normally distributed with zero mean and known variance. Moreover, suppose we can observe the position of the spacecraft, but that the observations are corrupted by additive white Gaussian noise. Our job is to produce the best estimate of both the position and the velocity.

We let $x_1(t)$ denote the position of the spacecraft, and let its velocity be $x_2(t)$. It is given that $x_1(0) = 0$, but that $x_2(0)$ is a Gaussian random variable with mean zero and a known variance, call it x_2^*. The dynamics of the process are

$$\frac{dx_1}{dt} = x_2,$$
$$\frac{dx_2}{dt} = 0.$$

The observations are given by

$$z(t) = x_1(t) + v(t),$$

where $v(t)$ is a Gaussian noise process with zero mean and (co)variance $r\delta(t - \tau)$, $r > 0$.

Note that in this problem there is no uncertainty in the dynamics because only the initial state x_0 is not known exactly. Therefore, the noise process $u(t)$ is identically zero. So our system dynamics and observation process are described by the quantities

$$F = \begin{pmatrix} 0 & 1 \\ 0 & 0 \end{pmatrix}, \qquad H = (1 \quad 0), \qquad R = (r),$$

$$Q = (0), \qquad G = (0), \qquad M = \begin{pmatrix} 0 & 0 \\ 0 & x_2^* \end{pmatrix}, \qquad x_2^* \geq 0.$$

The equations for the components of the matrix Riccati equation then work out to be

$$\frac{dP_{11}}{dt} = 2P_{12}(t) - P_{11}^2(t)/r, \qquad P_{11}(0) = 0,$$

$$\frac{dP_{12}}{dt} = P_{22}(t) - P_{11}(t)P_{12}(t)/r, \qquad P_{12}(0) = 0,$$

$$\frac{dP_{22}}{dt} = -P_{12}^2(t)/r, \qquad P_{22}(0) = x_2^*.$$

The optimal filter gain matrix $K(t)$ is then

$$K(t) = P(t)H'R^{-1} = \begin{pmatrix} P_{11}(t)/r \\ P_{12}(t)/r \end{pmatrix}.$$

This leads to the equations for the optimal estimates of the position and velocity of the spacecraft as

$$\frac{d\hat{x}_1}{dt} = \hat{x}_2(t) + (1/r)P_{11}(t)[z(t) - \hat{x}_1(t)], \qquad \hat{x}_1(0) = 0,$$

$$\frac{d\hat{x}_2}{dt} = (1/r)P_{12}(t)[z(t) - \hat{x}_1(t)], \qquad \hat{x}_2(0) = 0.$$

Although these equations may be difficult to solve analytically in closed form, they are ideally suited to process numerically because the data $z(t) = x_1(t) + v(t)$ come from our measurements. Of course, the quantities P_{11}, P_{12}, and P_{22} needed to compute the Kalman gain matrix $K(t)$ can be precomputed off-line, if we wish.

Although there is much more to say about the Kalman filter and its many properties and uses, we shall refrain from doing so here. The reader can find as much detail as anyone would want or need in the chapter references. Instead let's close our discussion of control theory by showing the connections between the Kalman filter and the LQ control problem given earlier.

Duality—Yet Again

If we define a dynamical system that is the *dual* (or adjoint) to the dynamics of the filtering system, we obtain

$$t^* = -t, \qquad F^* = F', \qquad G^* = H', \qquad H^* = G',$$

where the starred quantities are those for the dual system and the unstarred ones are those for the filtering problem. So the dual system has the dynamics

$$\frac{dx^*}{dt^*} = F^* x^*(t) + G^* u^*(t),$$
$$y^*(t) = H^* x^*(t).$$

We now form an LQ control problem from this by comparing the matrix Riccati equation for the earlier LQ control problem with the matrix Riccati equation given by the Kalman Filter Theorem. Again denoting the quantities from the LQ problem (with stars) and those from the filtering problem (without stars), we find

$$Q^* = Q' = Q, \qquad R^* = R' = R, \qquad M^* = M' = M,$$

relations that all follow from the symmetry of the matrices Q, R, and M.

From these relations we find that to construct a purely deterministic LQ control problem whose solution is *identical* to that of the stochastic linear filtering problem, we need only make the substitutions indicated in Table 3.1. So by these relations we can convert any linear filtering problem to a completely equivalent LQ control problem, and vice versa.

This filtering-control duality, together with the results of Chapter 2 on deterministic chaos, certainly cast a dark shadow of doubt upon the notion that Mother Nature has some kind of mystically intrinsic random component, at least insofar as the mathematical pictures we paint of her are concerned. For any mathematical picture containing inherently stochastic quantities, it now seems a good bet that we can construct an equally good picture that is purely deterministic. This philosophical

Table 3.1. The control-filtering duality.

Filtering	LQ Control
t	$-t$
F	F'
G	H'
H	G'
Q	Q
R	R
M	M

point seems to be an especially appropriate note on which to close our deliberations on matters control-theoretic.

CHAPTER

4

The Hahn-Banach Theorem

Functional Analysis

The Big Picture

The English poet, artist, and all-around mystic, William Blake, once wrote, "To generalize is to be an idiot. To particularize is the lone distinction of merit; general knowledges are those knowledges that idiots possess." Blake obviously didn't know any mathematicians. Or at least he didn't hold them in very high regard if he did know some, because generalization is the stock-in-trade of the mathematical enterprise. And nowhere is that more apparent than in the emergence of the area we now call *functional analysis.*

The fruitful generalization in mathematics often involves starting from a commonsense concept such as a point on a line. A mathematical framework is then developed within which the *particular* example of a point in space is seen to be just a very special case of a much broader structure, say a point in three-dimensional space. Further generalizations then show this new structure itself to be only a special case of an even broader framework, the notion of a point in a space of n dimensions. And so it goes, one generalization piled atop another, each element leading to a deeper understanding of how the original object fits into a bigger picture. Let's pursue this idea further in search of how functional analysis came into existence when the need for more general mathematical structures—noneuclidean geometries, groups, and dynamical systems—arose in the early part of the twentieth century.

Let's begin by considering one of the most basic geometrical objects encountered in daily life, the points constituting a flat plane. Such objects form the basis, for example, for the location of intersections in a road-traffic network and the positioning of property lines for surveyors and realtors. Since the time of Descartes, we have known how to associate a unique pair of real numbers (x, y) with every such point p, so that we can identify the *geometrical* object p with the *algebraic* entity (x, y), that is, $p = (x, y)$. By now this is all so familiar that we see nothing remarkable in it. But, in fact, it's quite a remarkable claim to assert that the set of real-world points $x^* \in E^2$ constituting, say, the Northern Hemisphere on the surface of the Earth (minus the North Pole P), are identical (technically, isomorphic) to the points $x \in R^2$ of a flat plane. The justification for this claim is shown in Figure 4.1.

If an ordered pair of numbers represents a point in the plane, what about an ordered set of three, four, five, or more numbers? Shouldn't

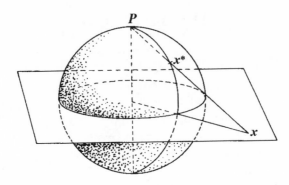

Figure 4.1. Isomorphism between the Northern Hemisphere and the plane.

these represent a point in increasingly higher-dimensional spaces? This indeed turns out to be the case. An ordered set of n real numbers, $(x_1, x_2, \ldots, x_n)$ is a perfectly faithful representation of a point in n-dimensional space, R^n, for any finite integer n. These numbers $p = \{x_i\}, i = 1, 2, \ldots, n$ are called the *coordinates* of the point p.

Given two points $p = (x_1, x_2, \ldots, x_n)$ and $q = (y_1, y_2, \ldots, y_n)$ in R^n, about the most basic question we can ask is whether p and q are the same point. If so, then the two sets of coordinates will coincide. But if not, then it's reasonable to wonder, how close are the two points to each other? The answer to this question is given by the famed Pythagorean Theorem, which states that the distance between p and q is given by the expression

$$d(p, q) = \sqrt{(x_1 - y_1)^2 + (x_2 - y_2)^2 + \cdots + (x_n - y_n)^2}$$

Note that this is not the only way to mathematically define the distance between two points. It is the most natural one, however, in the sense that it agrees with our usual notion of distance "as the crow flies."

But why stop at values of n that are finite? Couldn't we let $n \to \infty$? This is the question that David Hilbert asked in the early part of the twentieth century. The answer led directly to the field of functional analysis. Think, for a moment, about what a radical idea this is. What could it be like to be in an infinite-dimensional space? Finite-dimensional space means that there are a finite number of "directions" in which we can move from a given point to any other. Infinite-dimensional space, then, must mean that from any point there are an infinite number of different ways that we can go. Let's try to see what this might mean.

A good way to visualize what happens when n becomes infinitely large is to think of the set of coordinate values $(x_1, x_2, x_3, \ldots)$ defining a point in the infinite-dimensional space as constituting a rule, or *function*. For each position k in this infinite list of numbers, the rule simply assigns the value x_k at the kth location. Pictorially, this situation is shown in Figure 4.2, where the values x_k are shown as blocks of unit width for each $k = 1, 2, \ldots$.

This picture suggests that we could use a set of labels even bigger than the set of integers to build up an even bigger space of functions. So instead of the discrete sequence of numbers $(1, 2, \ldots)$ that lead to the set $(x_1, x_2, \ldots)$, we consider functions that assign a real number to every value of t in a continuous range. For example, the rule that assigns the value $x(t)$ to every positive real number t leads to the function $t : \ \rightarrow x(t)$, which can be plotted as a curve in the plane. In this case, we can regard the total area under the graph of the function $x(t)$, its *integral*, as the continuous analogue of the sum $x_1 + x_2 + x_3 + \cdots$. Symbolically, this area is denoted as $\int_0^\infty x(t)\, dt$.

By this artifice, we have managed to construct a continuous, rather than discrete, infinite-dimensional space whose "points" are now functions $p = x(t)$. If $q = y(t)$ is a second point of this space, we can now

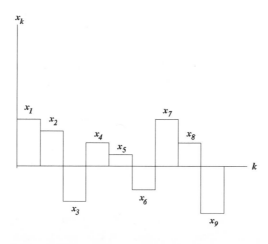

Figure 4.2. Histogram of an infinite sequence of numbers.

define the distance between p and q by the following generalization of the Pythagorean Theorem:

$$d(p, q) = \text{area under the graph of } (x(t) - y(t))^2$$
$$= \int_0^\infty (x(t) - y(t))^2 \, dt$$

The path we have outlined in going from points in a space such as a plane to points in an infinite-dimensional space of functions has led to a generalization of the idea of a space of points to a space of functions (*function space*), where each "point" in the space is now a function. The study of the properties of "functions of functions" was given the label *functional analysis* by the French mathematician Paul Lévy. The integral $I = \int_0^\infty x(s) \, ds$ of a function $x(t)$ is an example of a *functional* because it accepts a function $(x(t))$ and produces a real number (the value of the integral). But before taking up some of the basic elements of function spaces, let's pause to give a brief account of some classes of problems arising in the real world for which function spaces provide the right setting for their resolution.

Problems at Infinity

Around 800 B.C., Queen Dido was fleeing from the Phoenician city of Tyre, which was ruled by her tyrannical brother King Pygmalion. She arrived at the site of what was to become the city of Carthage and sought to buy land from the natives. They said that she could buy only as much ground as she could surround with a bull's hide. Dido accepted their terms, and cleverly cut a bull's hide into narrow strips that she tied together to form a single, very long strip. She then used her intuitive feeling that the maximum land area that could be surrounded would have the form of a circle, and proceeded to shape the strip into the circumference of a circle. The modern theory called the *calculus of variations* provides the mathematical underpinning confirming Queen Dido's intuition about a circle being the largest surface area that can be encompassed by a line of fixed length. Let's express Dido's problem in more formal terms, to see how function spaces enter crucially into its solution.

Suppose we are given a curve in the plane, whose x and y coordinates are given by

$$x = x(t), \qquad y = y(t), \qquad 0 \le t \le 1.$$

The curve is closed if $x(0) = x(1)$ and $y(0) = y(1)$, and there are no other intersections. The length of such a curve is then given by the functional

$$L = \int_0^1 \sqrt{\left(\frac{dx}{dt}\right)^2 + \left(\frac{dy}{dt}\right)^2}\, dt,$$

a number that depends on the functions $x(t)$ and $y(t)$. Similarly, the area A enclosed by such a curve is also a functional,

$$A = \int_0^1 \left(x\frac{dy}{dt} - y\frac{dx}{dt}\right) dt.$$

So Dido's problem can be stated compactly in the form: Over all closed curves of length L, find those that make the functional A as large as possible.

The set of all planar closed curves of fixed length constitutes an infinite-dimensional function space because there are an infinite number of ways to choose the functions x and y to generate independent curves. Dido was looking for a point in this space that would maximize the expression for the area. This is a problem in the field called the *calculus of variations,* which always involves function spaces because problems in this area involve finding a function that maximizes or minimizes some functional such as A. Tools developed in the calculus of variations show the answer to Dido's problem is indeed a circle. So if the myths of antiquity are to be believed, Dido's intuition to form the bull's hide cord into a circle of circumference L was the optimal way for her to deal with the stingy natives of Carthage. Let's look at some more problems in function space.

- *Game Theory:* One part of the mathematical theory of games considers finite, two-person, zero-sum games, in which each player has a finite set of actions that can be employed at each play of the game. The Minimax Theorem tells us the best strategies for each player. In particular, for two-person, zero-sum games, it asserts the existence of a distinguished point in a finite-dimensional space of discrete probability distributions, a point characterizing the optimal strategies for each player. The space of such distributions is an infinite-dimensional space, so the optimal strategies constitute a point in this space, just like the solution to the Dido problem. Let's consider now a more elaborate kind of game, called a *pursuit-evasion game,* in which infinite-dimensional spaces arise naturally. One of the simplest examples of such a game is the case of the homicidal chauffeur. It goes like this.

The action takes place in the plane (think of an infinitely large shopping mall parking lot). The chauffeur C moves at a fixed speed w, but the turning radius of his car is restricted. Thus, the radius of curvature of the car's motion is bounded by a finite quantity, call it R. So the chauffeur steers by selecting the value of the curvature at each moment t. Put another way, the chauffeur chooses a function $r(t)$ such that $|r(t)| \leq R$ at each moment t, where $r(t)$ is the curvature of the car's path at time t. The pedestrian P moves with simple motion. His speed v is fixed, and he chooses only his direction d of travel at each instant. So P must select a function $d(t)$. To make the problem interesting, we assume that the car has a greater speed than the pedestrian, so that $w > v$. But note that the pedestrian is more agile because abrupt changes in his direction of travel are allowed whereas the chauffeur's changes in direction are bounded by the curvature constraint. Capture occurs when the distance between C and P is less than some given quantity ϵ (e.g., when $\epsilon = 0$).

There are many questions that arise in the context of this homicidal chauffeur game. Probably the most basic is simply to ask, When can C catch P? Clearly, if the curvature constraint on the chauffeur is strong enough (large R), the radius of capture is small (small ϵ), and the two speeds don't differ greatly (w is not much larger than v), then the pedestrian P can always escape merely by sidestepping whenever the car gets close enough. This will always work because the limitation on C's curvature prohibits a sharp enough turn to effect capture. So the problem is to find precise conditions on the values of R, ϵ, w, and v that enable the chauffeur to run down the pedestrian.

From our point of view, what's interesting about this problem is that the chauffeur C must choose a function $r(t)$ and the pedestrian P a function $d(t)$, in such a way that C's choice minimizes the distance between them while P's choice maximizes it. Because these functions live in two different infinite-dimensional spaces, the homicidal chauffeur problem lies outside the bounds of traditional game theory; the solution is a point in a *product* of two infinite-dimensional spaces, requiring the tools and methods of functional analysis for its elucidation.

- *Atmospheric Radiative Transfer:* Consider a two-dimensional (technically, plane-parallel) slice of atmosphere of thickness T over a particular location on Earth, as shown in Figure 4.3. Here we see this atmospheric slab illuminated by parallel rays of solar radiation of net flux π entering the atmosphere at an angle whose cosine is u. As this radiation enters the atmosphere, it is scattered by interaction with particles in the slab. Our concern is with the source function, $J(v, t)$, which expresses the rate of production of scattered radiation emerging at an angle whose cosine is v at a height t above the ground inside the atmosphere. The function $J(v, t)$ is crucial for all work in radiative transfer because the all-important information about how much solar radiation is reflected out the top of the atmosphere in a direction v, $R(v, T)$, and how much is transmitted to the bottom, $\tau(v, T)$, are both special cases of the source function. Specifically, $R(v, T) = J(v, T)$ and $\tau(v, T) = J(v, 0)$. So in a certain sense, if you know the source function, you know everything there is to know about scattered radiation in an atmosphere.

If we assume that the scattering inside the slab is uniform in all directions (isotropic), and that the albedo, the number of particles produced when two particles interact, is c, a number less than 1, then the equation governing the source function is

$$J(v, t) = \frac{c}{4} \exp[-(T - t)/v] + \frac{c}{2} \int_0^t E_1(|t - t'|) J(v, t')\, dt'. \quad (*)$$

Here the quantity E_1 is the exponential integral function

$$E_1(|t - t'|) = \int_0^1 \exp(-|t - t'|/z) \frac{dz}{z}.$$

Expression $(*)$ is an *integral equation* for the unknown source function J. Technically, this is what is called a Fredholm integral equation (because the limits of integration are fixed between 0 and 1) of the second kind (because the unknown function J appears both inside and outside the integral sign). This type of integral equation is named after the Swedish mathematician Ivar Fredholm, who studied them in the early part of the twentieth century. The first term on the right-hand side of $(*)$ accounts for the radiation produced when a particle enters the atmosphere and has its first interaction at the depth t, whereas the second term measures

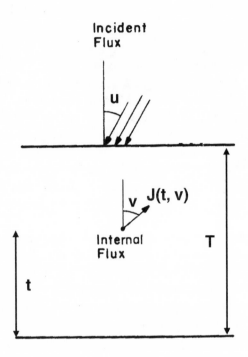

Figure 4.3. A planar atmospheric slab illuminated by parallel rays of solar radiation.

radiation produced from particles that are created at level t and then move to a level t' for their next interaction. Finally, the exponential integral function E_1 simply measures the likelihood that a particle produced at t will have its next interaction at t'.

The precise form of the equation for the source function J is not particularly important for us. What is important is that the unknown quantity J in the integral equation $(*)$ is a *function,* an element in an infinite-dimensional function space. So to assert that a solution of the integral equation even exists, we are going to have to use tools appropriate for finding elements in a function space. In this particular case, it turns out to be very useful to employ infinite-dimensional analogues of the fixed-point theorems discussed in Chapter 2 of *Five Golden Rules* to assure ourselves that there really exists a function J that solves the equation $(*)$. But more about that later in this chapter.

- *Quantum Mechanics:* Suppose we have an electron with energy E and mass m moving along a line on which there is a potential energy V. If the position of the electron on the line is denoted by t, then quantum theory tells us that the equation for the probability amplitude ψ of the electron's motion is given by the one-dimensional Schrödinger equation

$$\frac{\hbar}{2m}\frac{d^2\psi}{dt^2} + (E - V)\psi = 0. \qquad (**)$$

Here $\hbar$ is a fixed constant. Information about the likelihood of the electron's position, momentum, spin, or any other quantum attributes are given by performing various mathematical operations on the wave function ψ. So in a definite sense, the function $\psi(t)$ encapsulates all the information that we can ever hope to have about the various properties of the electron—at least it does if you believe in the conventional wisdom of quantum mechanics, the so-called Copenhagen interpretation. We consider this matter further later in this chapter. For now, simply observe that the solution ψ of the Schrödinger equation, even in this simplest of all settings, is still a point in a space of functions. In fact, if we assume the electron is confined to a region $0 \le t \le T$, so that the potential $V = 0$, equation $(**)$ has the explicit solution

$$\psi_n(t) = \sqrt{\frac{2}{T}} \sin\frac{n\pi t}{T}, \quad n = 1, 2, 3, \ldots,$$

where each $\psi_n(t)$ is an element of the space of trigonometric functions.

- *Distributed Control Systems:* Consider a copper metal rod of length 1 with insulated ends. Suppose we can add or remove heat at the midpoint of this rod, and that we measure the average temperature over the entire rod. Our goal is to apply heat or cold so as to keep the average temperature as close to zero as possible while using as little heat input as we can.

Mathematically, this control problem is similar to those considered in the last chapter—with the major exception that the space of states for this system is a function space, not a finite-dimensional space of points. Specifically, a state in this system is a function $u(x, t)$ giving the temperature at location x on the rod at time t. Symbolically, the equation governing the evolution of the state $u(x, t)$ is given by

$$\frac{\partial u}{\partial t} = \frac{\partial^2 u}{\partial x^2} + \delta(x - \tfrac{1}{2})v(t), \quad 0 \le x \le 1, \quad u(0, t) = u(1, t) = 0.$$

Here $\delta(\cdot)$ is the Dirac delta function, an infinitely high "spike" centered at $x = \frac{1}{2}$, with $x = 0$ elsewhere, and $v(t)$ is the control input function. This is an example of what is called a *distributed control process* because the control can be "distributed" over the entire spatial region $0 \le x \le 1$ (although in this elementary example it is concentrated at the single point $x = \frac{1}{2}$). Problems of this type, in which the state is governed by a partial differential equation such as the heat equation above, are common in engineering and physics.

Even more elementary are problems whose dynamics contain time-lag effects. A typical problem of this sort arises in business environments, when a decision taken at time t begins to affect the system only at some future time $t + \tau$, where $\tau > 0$ is the time lag. In this situation, the state of the system changes in accordance with the differential-delay equation

$$\frac{dx}{dt} = f(x(t), x(t - \tau)),$$
$$x(t) = g(t), \quad -\tau \le t \le 0.$$

So the natural state space of this problem is not a finite-dimensional set of points, but rather a function space whose properties are set by the nature of the initial function $g(t)$.

The foregoing list of problems from game theory, atmospheric physics, quantum mechanics, and engineering control theory gives ample testimony to the ubiquity of function spaces in applied mathematics. Functional analysis is therefore the proper branch of mathematics for the study of the properties of such spaces.

The Tale of Two Coffee Houses

In the days before World War II, there were two coffee houses in the Polish town of Lvov that served as a kind of "clubhouse" for the mathematicians centered around the University of Lvov. The first, Café Roma, was initially the more important of the two. It was there that mathematicians such as Stanislaw Ulam, Stefan Banach, Stanislaw Mazur, Mark Kac, and Hugo Steinhaus gathered after the weekly meetings of the local chapter of the Polish Mathematical Society. The meetings were usually held in a seminar room at the university, but the really fruitful discussions took place over a single cup of coffee or a large glass of tea, which could be nursed for several hours at the Café Roma, following the "official" meetings of the society.

Banach was by far the most prominent of the mathematicians who frequented the Café Roma, and he used to spend several hours—even days—there, especially toward the end of the month when the university salaries were paid. But one day Banach became annoyed with the credit situation at the Café Roma and decided to move to the cafe next door, a place called the Scottish Café. Now this cafe, a mere twenty yards down the street from the Roma, became the center of mathematics in Lvov.

One day, Banach decided that because so many ideas and problems were discussed at the Scottish Café, they should be entered into a book so they wouldn't be forgotten. So he bought a large, bound notebook and started to enter problems into it, the first entry bearing the date July 17, 1935. This notebook was kept at the Scottish Café by a waiter who was well-educated in the ritual of the mathematicians. When Banach came in, it was sufficient for him to say, "The book, please," and it would magically appear along with the cups of coffee. As the years passed, "the book" grew to be a collection of nearly 200 problems, most of which have by now been solved. As an example of one of these problems, here is problem 153, entered by Mazur on November 6, 1935. As with many of the problems in the book, it carried a reward for its solution—in this case a live goose!

> Given a continuous function $f(x, y)$ defined for $0 \leq x, y \leq 1$ and the number $\epsilon > 0$, do there exist numbers $a_1, \ldots, a_n; b_1, \ldots, b_n; c_1, \ldots, c_n$, such that $|f(x, y) - \sum_{k=1}^{n} c_k f(a_k, y) f(x, b_k)| \leq \epsilon$ for all x, y in the interval $0 \leq x, y \leq 1$?

This problem became famous as a forerunner of more general approximation problems. It asks whether the original function can be approximated arbitrarily closely by a new function, one generated by using values of the original function at a finite number of points. The answer to problem 153 is negative, as was shown by Per Enflo in 1972. Amusingly, the goose promised to the solver of the problem was presented to Enflo a year or so later following his lecture on the problem in Warsaw. (*Technical note:* The theorem is true if the function $f(x, y)$ has more structure than simply being continuous; for example, if f has a continuous first derivative in either x or y.)

As a historical document, the Scottish problem book is important because it traces the thinking of the group of mathematicians around Banach who were responsible for much of the development of functional analysis. Although the Germans, led by David Hilbert, made major contributions to the field in its infancy, it was really the impetus of the work done by the Lvov group that set the scene for what today is called functional analysis. So let's delve into some of the major conceptual issues and problems that were floating around the tables at the two Lvovian cafes before World War II turned everything upside down.

Spaces and Functionals

Consider again the two-dimensional plane consisting of all pairs of points (x, y), where x and y are real numbers. We have seen that the distance $d(p, q)$ between two such points $p = (x_1, y_1)$ and $q = (x_2, y_2)$ is given by the Pythagorean Theorem as

$$d(p, q) = +\sqrt{(x_1 - x_2)^2 + (y_1 - y_2)^2}. \tag{†}$$

The function $d: X \to R^+$ creates a nonnegative real number, the positive square root, for every pair of points in a space X (in this example, X is the plane, consisting of all points $p = (x, y)$, where x and y are real numbers). Such a function possessing the following three properties is called a *metric:*

$$d(p, q) = 0 \text{ if and only if } p = q \qquad \text{(reflexivity);}$$
$$d(p, q) = d(q, p) \qquad \text{(symmetry);}$$
$$d(p, q) \leq d(p, r) + d(r, q) \qquad \text{(triangle inequality).}$$

The particular metric (†) is the one with which we are most familiar from ordinary euclidean geometry. That is, it is the length of the straight line connecting p and q. But there are many other metrics that can be used to measure "distance" between points in the plane. For example, there is the "taxicab," or "Manhattan," metric

$$d(p, q) = |x_1 - x_2| + |y_1 - y_2|,$$

so called because it involves calculating distance by measuring only in the rectangular directions: up–down and right–left. In this metric, the shortest distance between two points p and q is not a straight line, but the distance that a taxi would have to drive to move from p to q in a rectangular road network such as that found in Manhattan.

You might wonder why having a means to measure distance between points in a space is important. The reason is that we often want to assert the existence of some point in the space as the solution of some equation or another, as, for example, in the Schrödinger equation or the heat equation discussed earlier. One of the most common ways to do this is to construct a sequence of points $\{p_0, p_1, p_2, \dots\}$ that converge to the point that solves our problem. This is a good way to find a solution because it's often rather easy to construct approximate solutions. So if we can get these "almost" solutions to converge, we will arrive at a true solution to the problem. But in order to talk about "convergence," we need to have some way of measuring how close one point is to another; hence, the need for something like a metric.

A set of points, together with a metric defined on the set, is called a *metric space*. The plane with either the euclidean or the taxicab metric is an example of such a space, but there are others. One that is very important for us is the set of all continuous functions defined on a closed interval, such as all points between 0 and 1. We denote this space as $C[0, 1]$. We can introduce a metric into this space by the rule:

$$d(p, q) = \max_{0 \le t \le 1} |p(t) - q(t)|,$$

where p and q are continuous functions on $[0, 1]$, as defined earlier.

Suppose we take the sequence of rational numbers generated by the following rule: the nth element of the sequence is defined to be $p_n = (1 + 1/n)^n, n = 1, 2, \dots$. Putting in the values $n = 1, 2$, and so forth, we construct the sequence of rational numbers

$$p_1 = 2, \quad p_2 = 9/4, \quad p_3 = 64/27, \dots.$$

But to our surprise, the limit of this sequence of rational numbers turns out to be the transcendental number e, the base of natural logarithms, which has the value $e = 2.71828182845\dots$, and is not a rational number at all. So here we have a metric space, the set of rational numbers, in which a sequence converges to an element that is not a member of the space. This is bad—we don't want to find solutions to equations that fail to belong to the set of objects possessing properties that we want the solution to have, such as continuity, boundedness, or measurability. So to rule out such situations, we introduce the notion of a *complete metric space*. Roughly speaking, a metric space is complete if every convergent sequence of elements in the space has as its limit another element of the space. (Technically, the sequences must be what are termed Cauchy sequences. But that's a fine point that we can skip over in a treatment of this sort.)

If we impose some conditions on the space of objects under consideration that demand that they satisfy the requirements for what is called a *vector space,* then the generalization of the idea of a space of real numbers, in which the usual rules of addition and multiplication of objects in the space by a constant can be defined. Then we can introduce another way to define the distance between points of the space. We won't go into the details needed to single out a vector space from a more general space of objects, other than to say that all the spaces we have talked about thus far—points in the plane, the space of continuous functions, and so on—are vector spaces. This new way of defining distance in a vector space involves imposing what's called a *norm* on the objects constituting the space. Essentially, the norm is a measure of the "size" of any point in the space. So, suppose we have a space of points X. A norm on X is given by a rule $\| \cdot \|$ that assigns a nonnegative real number to every element of X.

The Pythagorean metric we used earlier for the plane is an example of a norm on that space. In the space $C[0, 1]$ of continuous functions on the interval $[0, 1]$, the norm of a continuous function f is given by the rule:

$$\|f\| = \max_{0 \le t \le 1} |f(t)|.$$

If we have a space with a norm defined, then we can immediately induce a metric d on the space as well, by the simple rule:

$$d(x, y) = \|x - y\|.$$

Because the norm automatically gives rise to a metric by the aforementioned rule, it makes sense to speak about a complete normed space. This is a space in which every convergent sequence of points converges to an element of the space, and the metric used is the one induced by the norm. Such spaces are so important in mathematics that they have a special name. They are called *Banach spaces,* in honor of Stefan Banach, the Polish mathematician mentioned earlier in connection with the famed Scottish Café in Lvov. It was Banach's work (much of it done at the café) that laid the foundations for functional analysis. Now let's finally talk about the subject of functional analysis—functionals.

A *functional* f on a (real) space X is simply a rule that assigns a real number to each element of X. Here are some important examples.

1. *Inner Product.* If $X = R^n$, the space of ordered n-tuples of real numbers $x = (x_1, x_2, \ldots, x_n)$, and $a = (a_1, a_2, \ldots, a_n)$ is a fixed element of X, then

$$f(x) = \sum_{i=1}^{n} a_i x_i = a_1 x_1 + a_2 x_2 + \cdots + a_n x_n = < a, x >$$

is a functional on X.

2. *Integration.* Suppose $X = C[0, 1]$, the set of continuous functions on the interval $0 \le t \le 1$. Then the operation of calculating the area under the curve $g(t) \in C[0, 1]$ is a functional. It is given explicitly as the operation of integration

$$f(g) = \int_0^1 f(s) \, ds.$$

Both of these examples are illustrations of what is called a *linear* functional because the rules for assigning a real number to an object of the space are additive and homogeneous with respect to scalar multiplication. That is, in both cases $f(x + y) = f(x) + f(y)$ and $f(\alpha x) = \alpha x$, for x, y in X and any real number α. Here is an example of a nonlinear functional.

3. *Quadratic Forms.* Take $X = R^2$. Then

$$f(x) = \tfrac{1}{2}(x_1^2 + x_1 x_2 + x_2 x_1 + x_2^2),$$

is a nonlinear functional on X. It's easy to see that this rule can be extended to the space $X = R^n$ and generalized by defining

$$f(x) = \tfrac{1}{2} \sum_{i,j=1}^{n} a_{ij} x_i x_j, \qquad a_{ij} = a_{ji}. \tag{$\ddagger$}$$

Here the quantities a_{ij} are real numbers. So if $n = 2$ and $a_{ij} = 0$ (for $i \ne j$) and $a_{ii} = 1$, we recapture the original situation in two-dimensional space.

Expression (‡) is called a *quadratic form* on X and plays an important role in control and optimization theory, as we saw in *Five Golden Rules,* as well as in Chapter 3 of this volume.

The first example above, introducing the notion of an inner product, opens that way to our last important type of space. This is what's called a *Hilbert space,* named after the great German mathematician, David Hilbert, who introduced this kind of space in the 1920s. Suppose we consider the real vector space R^n, and define the *inner product* of two elements x and y of R^n by the rule:

$$\langle x, y \rangle = \sum_{i=1}^{n} x_i y_i.$$

This rule assigns a real number to every pair of elements of R^n. It's easy to see that the inner product $\langle \cdot, \cdot \rangle$ gives rise to a norm on R^n in the following way:

$$\|x\| = \sqrt{x_1^2 + x_2^2 + \cdots + x_n^2} = \sqrt{\langle x, x \rangle}.$$

A *Hilbert space* is simply a vector space on which we can define an inner product, such that the space is complete using the norm derived from the inner product, as shown above. All the finite-dimensional spaces R^n are examples of Hilbert spaces, as is the space of square-integrable functions $L_2[0, 1]$, which consists of functions $g(t)$ defined on the interval $0 \le t \le 1$ whose integral is finite, namely,

$$\int_0^1 g^2(s) \, ds < \infty.$$

In this latter case, the inner product is defined for two square-integrable functions g and h by the expression

$$\langle g, h \rangle = \int_0^1 g(s)h(s) \, ds.$$

The classical Pythagorean Theorem is such a result about the length of the long side of a triangle whose other sides are orthogonal to each other. The inner product allows us to define abstractly the concept of orthogonality of two vectors in a Hilbert space. The definition is simple: Two points p and q of a Hilbert space are orthogonal if and only if their inner product vanishes, $\langle p, q \rangle = 0$. This follows from the fact that the inner product generalizes the idea of the angle between two lines and is set up so that if the angle is 90 degrees, the inner product vanishes.

Orthogonality in a Hilbert space $\mathcal{H}$ is important for the following reason. Suppose $\mathcal{M}$ is a closed subspace of the overall space $\mathcal{H}$, and that we let $\mathcal{M}^\perp$ be the set of all points in $\mathcal{H}$ that are orthogonal to every element of $\mathcal{M}$. Then every point x in $\mathcal{H}$ can be expressed as the sum $x = m + m^\perp$, where m is a point in the subspace $\mathcal{M}$ and $m^\perp$ is a point in the space $\mathcal{M}^\perp$. This decomposition result enables us to easily solve many optimization problems involving finding a particular element of $\mathcal{H}$, as we'll see in just a moment.

We have now discussed the three principal types of spaces beloved of functional analysts everywhere—metric spaces, Banach spaces, and Hilbert spaces. The main feature distinguishing one from the other is the way in which the distance between two points is measured in the space. Table 4.1 summarizes the situation.

Table 4.1. Types of spaces and their distance measures.

Type of Space	Distance between Points p and q
Metric space	Metric $d(p, q)$
Banach space	Norm $\|p - q\|$
Hilbert space	Inner product $\langle p - q, p - q \rangle$

With these ideas of spaces and functionals at our disposal, let's look at some ways in which they can be employed to help answer important applied questions.

Minimum Distances and Maximum Profits

A large number of problems in engineering and economics can be expressed as finding an element of a Hilbert space that is as small as possible, that is, it has minimum norm. This is because we often want to minimize energy or costs or time, and these quantities are best measured in a Hilbert space setting. Here's an example.

In Chapter 3 we considered control problems for which the optimization criterion was a quadratic function of the state and control, and the side constraint linking the state and control was a linear differential equation. A very simplified problem of this sort is the passive RC (resistance-capacitance) electrical circuit shown in Figure 4.4, which consists of a voltage source v, a current i, a capacitor of capacitance c, and a resistor of resistance r.

The equation governing the change of the voltage $v(t)$ is

$$\frac{dv}{dt} = \frac{1}{c}i(t), \quad v(0) = v_0.$$

The energy dissipated in the resistor over the time interval $[0, T]$ is

$$J = \int_0^T ri^2(s)\,ds.$$

Suppose we want to control the current $i(t)$ so that the final voltage is fixed at a level $V(T) = v^*$, while minimizing the energy J used to generate this voltage. Let's see how the solution to this type of control process can be neatly expressed as a problem in Hilbert space.

To keep things as simple as possible, consider the problem of minimizing the scalar quadratic cost function

$$J = \int_0^T [x^2(t) + u^2(t)]\,dt,$$

where x and u are related by the linear differential equation

$$\frac{dx}{dt} = u(t), \quad x(0) = c. \tag{$\P$}$$

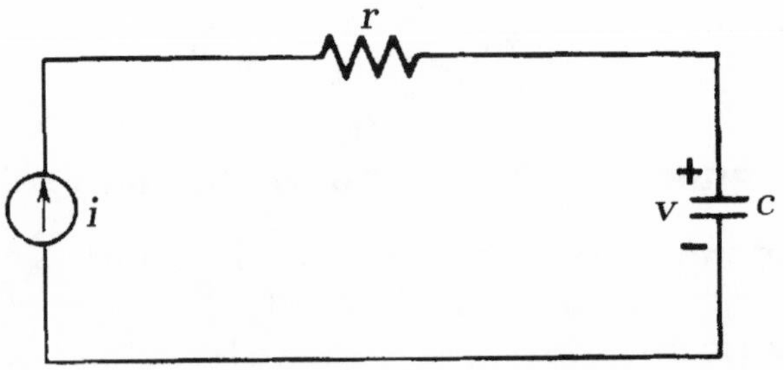

Figure 4.4. An RC electrical network.

This type of problem arises whenever we want to reduce the state x to as small a value as possible quickly, while at the same time using as little control resource u as possible. So, for instance, in the electrical circuit example we want to drive the voltage to v^* as quickly as we can. This can be represented as minimizing the difference $V(T) - v^*$. But we also don't want to use a lot of energy in the process. The quadratic criterion involving the squares of x and v is a good compromise between these conflicting objectives, and can even be used to weight the relative importance of the two goals by assigning a numerical coefficient to either the state or the control term in the integral J. But for simplicity and ease of writing, we have assumed here that both goals are equally important.

To formulate this problem as one in Hilbert space, we use elementary calculus to rewrite the differential equation (¶) in integral form as

$$x(t) = c + \int_0^t u(s)\, ds.$$

Since both the state function $x(t)$ and the control function $u(t)$ must be square-integrable for the objective function J to be defined, it's natural to formulate our problem in the Hilbert space

$$\mathcal{H} = L_2[0, T] \times L_2[0, T],$$

consisting of ordered pairs of functions (x, u) whose squares are integrable over the interval $[0, T]$. This space can be given an inner product defined by

$$\langle (x_1, u_1), (x_2, u_2) \rangle = \int_0^T [x_1(s)x_2(s) + u_1(s)u_2(s)]\, ds.$$

This inner product induces the norm

$$\| \langle x, u \rangle \|^2 = \int_0^T [x^2(s) + u^2(s)]\, ds.$$

The set of pairs (x, u) satisfying the constraint (¶) is called a *linear variety* in $\mathcal{H}$. Let's call this variety V. Then the control problem becomes to find the element of V of minimum norm. Happily, there is a result for Hilbert space that tells us exactly how to find such an element.

PROJECTION THEOREM FOR LINEAR VARIETIES *Let $\mathcal{H}$ be a Hilbert space and V be a closed subspace of $\mathcal{H}$. Corresponding to any element* h *in $\mathcal{H}$, there is a unique element* v_0 *in V such that* $\|h - v_0\| \leq \|h - v\|$ *for all* v *in V. Moreover, a necessary and sufficient condition that* v_0 *be the unique minimizing element is that* h $-$ v_0 *be orthogonal to V.* ∎

What this says is that if we are given an element h of the "big" space $\mathcal{H}$ and want to find the element in the subspace V that is closest to h (where "closest" is measured by the norm), then we need only look for the element v_0 that is the orthogonal projection of the element h onto the subspace V. That is, the element v_0 is such that $h - v_0$ is orthogonal to the subspace V. Figure 4.5 gives a geometric illustration of this fundamental result.

Applying this result to our control problem, all we need to show to ensure that a unique solution $u^*(t)$ exists is to prove that the linear variety V is closed. This follows by a simple argument using the triangle inequality in the Hilbert space $\mathcal{H}$.

It turns out that these Hilbert space results involving minimal norms and the linear variety V can be generalized to nonlinear varieties of a special kind: convex sets. Informally, a *convex set* is one in which all

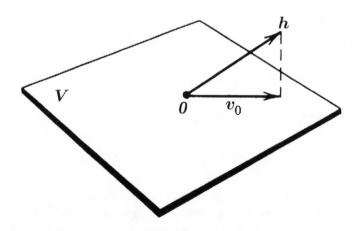

Figure 4.5. The Projection Theorem for linear varieties.

the points on every straight line joining any two points in the set are also in the set. So, for instance, a circle in the plane is a convex set, as is a triangle, but an annulus is not.

The basic result generalizing the classical Projection Theorem for linear varieties is the following:

PROJECTION THEOREM FOR CONVEX SETS *Suppose* h *is an element of a Hilbert space* $\mathcal{H}$, *and* K *is a closed convex subset of* $\mathcal{H}$. *Then there is a unique element* k_0 *in* K *such that*

$$\|h - k_0\| \le \|h - k\|$$

for all k *in* K. *Moreover, a necessary and sufficient condition that* k_0 *be the unique minimizing element is that* $\langle h - k_0, k - k_0 \rangle \le 0$ *for all* k *in* K. ■

As with the case for linear varieties, there is a geometrical interpretation of this theorem, too, as shown in Figure 4.6. Let's see how to use this extended version of the Projection Theorem to solve an important type of approximation problem.

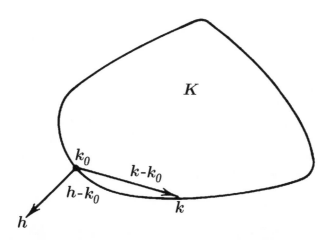

Figure 4.6. The Projection Theorem for convex sets.

Suppose you have made measurements on some financial data, such as the Dow Jones Industrial Average (DJIA). Let's represent this data in a vector of numbers $h = (h_1, h_2, \ldots, h_n)$, where the ith component of h is the DJIA at time period i. Now, suppose you also have measurements on other financial variables, such as the prime interest rate, the unemployment rate, the consumer confidence level, and the consumer price index. We represent this second set of data by vectors $y^1, y^2, \ldots, y^n$, where, again, the tth component of the vector y^i is the value of the ith variable at time period t. So, for instance, y_t^1 may be the interest rate at time t, whereas y_t^2 could be the unemployment rate, and so on. It's reasonable to suppose that we have only a finite number of measurements, so that $j = 1, 2, \ldots, N$. So the vectors h and y^i ($i = 1, 2, \ldots, n$) are each elements of the finite-dimensional space R^N. The problem is how to best approximate the DJIA, the vector h, by a linear combination of the other variables $y^1, y^2, \ldots, y^n$. Symbolically, we want to find real numbers $a_1, a_2, \ldots, a_n$ so that

$$\|h - (a_1 y^1 + a_2 y^2 + \cdots + a_n y^n)\|$$

is as small as possible. Often, in practice, we ask that the approximation coefficients all be nonnegative because these numbers frequently have a physical interpretation, such as a rate of flow, which doesn't make sense if the number is negative. So to make things interesting, let's impose this condition on the problem, as well.

Mathematically, we can formulate this approximation problem as that of finding the minimum distance from the point h to the convex cone

$$K = \{y \text{ such that } y = a_1 y^1 + a_2 y^2 + \cdots + a_n y^n, a_i \geq 0 \text{ for all } i\}.$$

This convex set is easily seen to be closed, so by the Projection Theorem for convex sets there is a unique minimizing set of coefficients $a_1^*, a_2^*, \ldots, a_n^*$. These coefficients give rise to the optimal approximating element $h^* = a_1^* y^1 + a_2^* y^2 + \cdots + a_n^* y^n$. According to the theorem, this element must satisfy the condition

$$\langle h - h^*, y - h^* \rangle \leq 0, \quad \text{for all elements } y \text{ in } K.$$

A bit of algebraic manipulation leads to the simpler condition

$$\langle h - h^*, y^i \rangle \leq 0, \quad \text{for all } i = 1, 2, \ldots, n,$$

with equality if $a_i^* > 0$. In other words, the optimal element h^* is "semi-orthogonal" to each data element y_i, and strictly orthogonal if the optimal coefficient is strictly positive. Unfortunately, the equations that follow from this result for actually determining the optimal coefficients $\{a_i^*\}$ are nonlinear, which makes them tricky to solve. But the simple functional-analytic results just given ensure that this set of equations has a solution—and it is unique.

In the universe of all spaces of mathematical objects, Hilbert spaces are very special; the ability to define an inner product on a space leading to a norm imposes a level of structure that leads to very detailed—and useful—results, such as the Projection Theorems. We'll see even more later in this chapter. But most of the infinite-dimensional spaces that we encounter in applied mathematics do not have such a nice, convenient structure. Yet we want to answer the same sorts of questions for objects in these spaces as we do for objects in Hilbert space. This raises the question of the degree to which results of the sort we have just seen for Hilbert spaces can be extended to more general situations. In particular, can we find a way to extend the Projection Theorems to the much broader class of Banach spaces, in which there is a norm, but not one derived from an inner product? If we can do this, then we'll have a tool for solving optimization problems in a much broader class of situations than when we have to use quadratic costs generated by an inner product in a Hilbert space. This is where the Hahn-Banach Theorem, perhaps the most central result in functional analysis, begins to enter our story.

The Other Side of a Space

A powerful weapon in the mathematician's arsenal is the idea of duality, which offers one of the few ways I know to get something for nothing. Basically, duality enables us to prove one theorem and, lo and behold, get a second theorem almost free of charge. To see how this works, consider the plain, old plane geometry of Euclid. There, the subject matter involves the relationship between points and lines in the plane. So, for instance, a typical theorem is the well-known result: "Every two distinct points in the plane determine a line." Now suppose you interchange the words "point" and "line" in this statement. The new statement is then, "Every two distinct lines in the plane determine a point." Miraculously, this dual statement is also a true fact about the relationship between points and lines in the plane. A few more experiments along these lines leads one to conjecture that *every* provable statement about points and lines has an equally valid dual statement, formed simply by interchanging the words "point" and "line." In short, there is a duality between points and lines in the plane. Indeed, this does turn out to be the case. Even more interestingly is that this notion of duality is not a peculiarity of euclidean geometry, but pervades almost every nook and cranny in mathematics. We saw an illustration of this in Chapter 3 on control theory, where the concepts of reachability and observability turned out to be dual to each other; every result on the reachability of a control system was at the same time a result about observability of a dual control system. Let's see how this same idea works in functional analysis.

Our focus to this juncture has been primarily with spaces of mathematical objects whose elements, for example, are continuous functions on an interval (the space $C[0, 1]$) or finite lists of real numbers arrayed as vectors (the space R^n, for some finite integer n). So if these objects are the "points," what are the "lines"? A big part of the genius of people such as Banach, the Austrian Hans Hahn, and other developers of functional analysis was the ability to answer such a question. The right answer is not entirely obvious because it's a lot more difficult to visualize a continuous function as a "point" than it is to visualize a point on a piece of paper.

As it turns out, the key to answering the question of what constitutes a "line" in a function space is to make use of the structure that glues the points of the space together. What this structure amounts to is being able to define an operation of "addition" in the space that "adds" two elements together to yield another element of the space. In such function spaces, the right notion of what constitutes the "lines" turns out to be the set of all linear functionals that can be defined on the space, and the "points" are the functions making up the space. Let's illustrate this with an example.

Consider the space R^n, consisting of vectors of n real numbers. A typical element of this space is written as the vertical array

$$x = \begin{pmatrix} x_1 \\ x_2 \\ \vdots \\ x_n \end{pmatrix},$$

where each x_i is a number. Probably the simplest linear functional that can be defined on this space is the one that just picks out the number in one of the positions in the array, say position k. This is sometimes called the "evaluation" functional. If we call this functional f_k, the action is $f_k(x) = x_k$.

As a finite-dimensional vector space, R^n is also a Hilbert space. This means that an inner product structure can be defined on this space. Define the vector e_i to be the element whose ith entry is 1, with all other entries being 0. Then the inner product of e_i with an element x in R^n is simply $\langle x, e_i \rangle = x_i$. In other words, this inner product just picks out the ith entry in x. So the seemingly more general linear functional f_k defined earlier can be *represented* in its action by the very special inner product operation $\langle \cdot, \cdot \rangle$ using the vectors e_j, $j = 1, 2, \ldots, n$. This is a special case of a much more general theorem from functional analysis due to the Hungarian mathematician Frigyes Riesz.

THE RIESZ REPRESENTATION THEOREM (WEAK FORM) *Given a linear functional* f *on the space* R^n, *there exists a vector* ℓ *in* R^n *such that* f(x) = $\langle x, \ell \rangle$. ∎

This is a very useful result. There are in principle an almost endless variety of linear functionals that one might imagine defining on the space R^n. Riesz's theorem says that as far as their end result is concerned, the actual number that one of these functionals produces when applied to an element of R^n can be represented by a very special type of functional—an inner product. So in this sense, there is really only one type of functional on R^n, and that type is the inner product. Of course, this tells us nothing about the *structure* of other sorts of functionals, such as the evaluation functional discussed above. That functional, for example, certainly has a different mathematical structure than an inner product. But what the Riesz Representation Theorem says is that if it's the action of the functional that you care about, then you don't need to know anything about the structure. Just find the appropriate element ℓ, then use the inner product and you will recreate the action of whatever functional you're studying.

For technical reasons, it's convenient to write the vector ℓ whose existence is guaranteed by the Riesz Representation Theorem as a *row* of n numbers, while continuing to write elements x in R^n as columns. The set of all such row vectors ℓ is then called the *dual space* to R^n, and is written as $(R^n)^*$. This construction is not confined to R^n; it is perfectly general. Given a vector space X (of "points," finite- or infinite-dimensional), the dual space X^* consists of all linear functionals ("lines") that can be defined on X. In the special case of Hilbert spaces, these linear functionals can all be represented by an inner product, as the Riesz Representation Theorem shows. Of course, for general vector spaces the functionals are not representable in this form because there is usually no inner product structure possible on these spaces. But a more general form of the theorem exists, which extends the Hilbert space result to a wider class of situations. The technicalities of how this extension goes for more general function spaces such as $C[0, 1]$ are beyond the bounds of our introductory account here, but it can be done. The interested reader will find these constructions by consulting the material cited in the reference section for this chapter.

As with all the duality results in mathematics, the dual space structure we have just introduced allows us to prove results in the space X^* of linear functionals simply by first proving them in the space X—or vice versa. So, for example, if $X = C[0, 1]$, the set of continuous functions on the interval $[0, 1]$, the space dual to this turns out to be $X^* = BV[0, 1]$, the set of functions having bounded variation on $[0, 1]$. Consequently, for every result we can prove about continuous functions, such as the fact that a continuous function assumes minimum and maximum values on the closed interval $[0, 1]$, there is a corresponding result for functions of bounded variation. In short, from a mathematical point of view, the two spaces X and X^* are equal. We return to this notion of duality later. For now, let's pause to discuss what is probably the most important single result in all of functional analysis, the famed Hahn-Banach Theorem.

But Does It Exist?

In the mid 1920s, Hans Hahn, professor of mathematics at the University of Vienna, began the creation of a theory of linear functionals defined on *abstract* normed vector spaces. It's important to realize that prior to this time, mathematicians had focused their attention on *concrete* spaces of this sort, such as R^n and $C[0, 1]$, defining norms by various explicit algebraic formulas. Hahn, on the other hand, saw that these efforts were all special cases of a single, albeit far more abstract, setting: the normed vector space. The problem he faced was that it was far from clear that bounded linear functionals even exist for every possible normed vector space. So Hahn's problem was to ensure that his theory of functionals on a vector space was not vacuous; that objects such as bounded linear functionals really did exist for every possible normed vector space. Establishing this as a fact was the crucial breakthrough enabling the development of functional analysis. Hahn published his result on the existence of linear functionals on abstract spaces in 1927. Two years later, Stefan Banach independently discovered the same result, which today is rightly termed the *Hahn-Banach Theorem*. Like most great mathematical results, it comes in several different forms, each of which illuminates a different aspect of its nature. Here we will give two forms of this basic result, one algebraic and the other geometric. But first, a couple of preliminaries.

The dual space of a normed vector space is the set of bounded linear functionals on that space, so it's reasonable to suppose that we can define a norm on this space of functionals as well, so as to measure the "size" of the functional. This is because the dual structure mimics the structure of the primal space perfectly. If the primal space has a norm, then so should the dual space—and it does. Initially, it may seem that the right measure of size for a functional f at a point x from a space X is just the magnitude of the number the functional produces, $|f(x)|$ when applied to x. The norm might then be defined as the largest such number taken over all elements x in X. This is almost the correct definition of the norm of f, except that it places too much emphasis on the points x and not enough on the functional f. After all, the norm of f should really be a property of f and shouldn't be distorted by the size of the elements from X. So, for instance, the norm of the integration functional $\int_0^1 (\cdot) \, ds$ on the space $C[0, 1]$, which measures the area under the curve of the continuous function g from the space, should not be large just because the elements of the space happen to be "big" functions.

The easiest and most natural way to cure this difficulty is to divide the number $|f(x)|$ by the norm of the element x. This procedure then ensures that a "large" x will be normalized by its own magnitude when we look for the biggest value produced by f. Summarizing the situation, for the norm of a functional f we simply compute its value on each element of X, normalizing the value received in each case by the magnitude of x, and then take the largest of these numbers. Symbolically,

$$\|f\| = \sup_{\|x\| \leq 1} |f(x)|.$$

Here the supremum operation sup is used instead of the maximum, to account for the fact that the space X may not be closed, in which case the maximum of $|f(x)|$ may not actually be assumed by any element of X but only closely approximated. This is a technical point that need not concern us here because all the spaces we consider in this chapter are closed, allowing us to replace the sup operation by max.

The second preliminary item we need before stating the Hahn-Banach Theorem is the idea of an extension of a functional. Suppose X is a normed vector space and that G is a subspace of X. Let f be a linear functional defined only on the elements of the subspace G. Then a linear functional F defined on X is called an *extension* of f if $f(x) = F(x)$ for all elements x in G. So an extension of f is simply a functional F that agrees with f on the subspace G. Clearly, there may be many extensions of a given functional f. We can also define the norm of f on G as:

$$\|f\|_G = \sup_{x \in G, \|x\| \leq 1} |f(x)|.$$

Now for one version of the Hahn-Banach Theorem.

___

HAHN-BANACH THEOREM—EXTENSION FORM *Every linear functional defined on a subspace* G *of a vector space* X *can be extended into the whole space* X *with preservation of norm. In other words, there exists a linear functional* F(x) *defined on all of* X, *such that*

1. F(x) = f(x) *for all* x *in* G *and*
2. ||f||G = ||F||x ∎

By this result we see that if there exists a functional on a subspace of a normed vector space X, we can generate from it a functional on the entire space. Consequently, the field of functional analysis is not vacuous if for every space X we can find a subspace of X on which a linear functional can be defined. Here's how to do this.

Let x_0 be an element of a given normed vector space X. Now, consider the set of elements $\{tx_0\}$, where t is some real number. In other words, create a collection of new elements of X by multiplying x_0 by the reals. The set $G = \{tx_0\}$ constitutes a subspace of X generated by x_0. Now define the functional $\phi(x) = t\|x_0\|$, for $x = tx_0$. The functional ϕ has the properties that it is linear, bounded, and not identically zero. Furthermore, $\phi(x_0) = \|x_0\|$ and $|\phi(x)| = |t|\|x_0\| = \|tx_0\| = \|x\|$, so that $\|\phi\| = 1$. Thus, by the Hahn-Banach Theorem, the functional ϕ can be extended to a bounded, linear functional on the entire space X. This is just what we need to guarantee that the study of linear functionals on X is not an empty exercise; there really do exist nontrivial linear functionals on every normed vector space.

Perhaps the most profitable way to think of the Hahn-Banach Theorem is to view it as an existence result for an optimization problem. Given a functional f on a subspace G of a normed space X, it's usually not difficult to extend f to the entire space. But, in general, the extension will be unbounded or at least have norm greater than the norm of f on G. Therefore, we pose the problem of finding an extension of minimum norm. The Hahn-Banach Theorem guarantees the existence of a solution to this problem, and even tells us the norm of the best extension (but, like most existence theorems, it fails to provide a prescription for how to actually find the minimal norm extension).

There are several duality principles in optimization theory that relate a problem stated in terms of points in a vector space to a problem expressed in terms of hyperplanes in the space. Many of these principles rest on the geometric relation shown in Figure 4.7. The shortest distance from a point x to a convex set K equals the maximum distance from the point to a hyperplane P separating the point from the convex set. The original minimization problem over vectors is then converted to a maximization problem over hyperplanes.

HAHN-BANACH THEOREM—GEOMETRIC FORM   *Given a closed, nonempty convex set* K *in a real vector space* X, *and a point* x *in* X *not in* K, *there exists a closed hyperplane* P *that separates* x *and* K. ∎

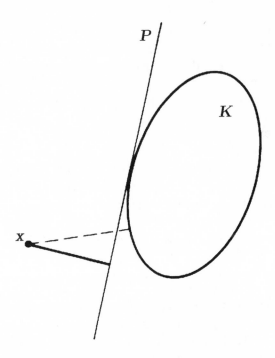

Figure 4.7. Duality between points and hyperplanes.

The connection between this geometric view and the earlier version of the Hahn-Banach Theorem is evident when we recognize that every hyperplane P in X is defined by a linear functional f. P is simply the set of points x such that $f(x) = c$, where c is some fixed real number. So there is no loss of generality in identifying P with the functional f. But because every functional f in the dual space X^* of X generates such a hyperplane, there is no mathematical difference between X^* and the set of all hyperplanes in X; they are isomorphic. So a more precise way of stating the geometric version of the Hahn-Banach Theorem is to say that given a closed, convex set K and a point x_0 not in K, there exists a linear functional f and a number c, such that x_0 lies in the set of points x in X such that $f(x) < c$, whereas the elements of K lie in the set of points x such that $f(x) > c$.

We have now spent a goodly number of pages developing background results that are probably of more interest to mathematicians than to practitioners. So it's natural to wonder what all this discussion of normed spaces, inner products, convex sets, representation theorems, and the like has to do with solving problems. What does the Hahn-Banach Theorem and its many progeny have to do with the world at large? It's now time to start delivering on the promise that functional analysis matters—and not just to mathematicians.

Big-Time Operators

Legend has it that as the American writer Gertrude Stein lay on her deathbed, one of her followers, grasping for one last tidbit of wisdom from this vast mother figure, asked, "Gertrude, Gertrude, what is the answer?" Gathering all her strength, Gertrude Stein rose up and replied, "What is the question?" Well, in mathematics there is no shortage of questions. The questions we care about are often stated in the form, What is the solution to such and such an equation? Among the most powerful weapons in the mathematical arsenal for answering this type of question are methods having their roots in functional analysis. Here we look at a few of them.

To this point in our narrative, the focus has been on functionals—transformations that map elements from a vector space X to the real numbers. To study the solutions of equations, we need to up the ante somewhat and consider a broader class of transformations, what are usually called *operators*. These are simply transformations from one abstract vector space X to another such set Y. So an operator A transforms an element of a normed vector space such as the set of continuous functions not to a number, like a functional would do, but to another function. Here is an example.

Suppose we consider the set $X = C^\infty[0, 1]$ of *smooth functions* defined on the interval $0 \leq t \leq 1$. These are functions that can be expanded in a Taylor series around every point in the interval, so that we can write such an object as

$$f(t) = a_0 + a_1 t + a_2 t^2 + a_3 t^3 + \cdots,$$

where the a_i are real numbers. Now let the operator A be defined on X by the rule

$$g = A(f) = a_1 + 2a_2 t + 3a_3 t^2 + 4a_4 t^3 + \cdots$$

Here the ith term in the function g is $ia_i t^{i-1}$. The function g is also a member of the set X of smooth functions, so that the operator A

transforms elements of X to itself. It's not difficult to see that A is also a *linear* operator, since $A(f_1 + f_2) = A(f_1) + A(f_2)$ and $A(\alpha f) = \alpha A(f)$ for every real number α. (*Technical note:* This operator A is called the "differentiation" operator because it maps a differentiable function f to its derivative, $df(t)/dt$.) Some additional examples of linear operators are the following:

1. *Matrices*—Let X be the space R^n of column vectors of n real numbers, and let $\{a_{ij}, i, j = 1, 2, \ldots, n\}$ consist of n^2 real numbers. Then the equations

$$b_i = \sum_{k=1}^{n} a_{ik} x_k, \quad i = 1, 2, \ldots, n \qquad (\S)$$

   define an operator A that transforms an element $x = (x_1, x_2, \ldots, x_n)$ of the space R^n into an element $b = (b_1, b_2, \ldots, b_n)$ of the same space. The operator A is linear and can be represented by the array

$$A = \begin{pmatrix} a_{11} & a_{12} & \cdots & a_{1n} \\ a_{21} & a_{22} & \cdots & a_{2n} \\ \vdots & \vdots & \ddots & \vdots \\ a_{n1} & a_{n2} & \cdots & a_{nn} \end{pmatrix}$$

   Such an array is called an $n \times n$ real *matrix*.

2. *Integral Operators*—Let $X = C[0, 1]$ and suppose that $K(x, y)$ is a function continuous on the square $0 \le x, y \le 1$. Set

$$g(t) = K(x) = \int_0^1 K(t, s) x(s) \, ds.$$

   Then it's easy to show that the function g also belongs to $C[0, 1]$, so that the operator K transforms X to itself. The operator K is a linear *integral operator,* and in fact, if we do the limiting operation carefully, turns out to be the limit, as $n \to \infty$, of the operator A defined in example 1.

Now let's see how this notion of an operator can be used to establish the existence of solutions to equations. Consider first the set of linear algebraic equations ($\S$) in example 1. Using the operator A defined by the $n \times n$ matrix above, we can compactly write these equations as

$$Ax = b.$$

If A, x, and b were real numbers, we could immediately solve this equation for x by setting $x = b/A$, or, more properly, $x = A^{-1}b$. This argument needs to be "souped up" a bit to work in the n-dimensional setting we're considering here, by finding an operator A^{-1} that is *inverse* to A. This can be done. But to prove the existence of the solution x to the equation $Ax = b$, let's follow another path, to illustrate a functional-analytic tool of a far more general character, the so-called *Contraction Mapping Theorem*.

Suppose X is a complete, metric space. Recall that this means there is a distance function $d(x, y)$ defined on every pair of elements of X, and that every convergent sequence of points of X converges to an element of the space. Furthermore, let A be an operator transforming a point of X to another point of X. Note that here we do not assume that A is linear. Now we can state the Contraction Mapping Theorem.

CONTRACTION MAPPING THEOREM   *Let* X *be a complete, metric space and* A *an operator on* X *with the following property:*

$$d(A(x), A(y)) \leq \alpha d(x, y),$$

for $0 < \alpha < 1$. *Then there exists exactly one point* x* *in* X *such that* A(x*) = x*.   ∎

(*Note:* The perceptive reader will note that the point x^* is a *fixed point* of the transformation A, as discussed in Chapter 2 of *Five Golden Rules*.)

This result is important for two reasons. First, it asserts conditions under which a general equation $A(x) = b$ has a solution in the space X for nonlinear operators A satisfying the "shrinking" condition of the theorem. Second, it provides a constructive procedure for actually finding such a solution. Here's how.

Suppose A satisfies the conditions of the Contraction Mapping Theorem, and we want to solve the equation $x = b + A(x)$. We construct a sequence of elements in X by the following rule: $x_0 = b$, $x_n = b + A(x_{n-1})$, $n = 1, 2, \ldots$. By the shrinking property of A, this sequence can be shown to converge to an element x^*, such that $x^* = b + A(x^*)$. But because X is a complete space, x^* must belong to it. So, by the Contraction Mapping Theorem, x^* is *the* solution to the equation.

- *Kepler's Equation in Celestial Mechanics:* As a simple example of the application of this result, consider Kepler's equation

$$x - a \sin x = b,$$

for the unknown x, where a and b are parameters such that $0 \leq a < 1$ and b is some finite real number. The function $f(x) = b + a \sin x$ satisfies the condition

$$|f(x_1) - f(x_2)| \leq \alpha |x_1 - x_2|,$$

for $0 < \alpha < 1$, so that f is a contraction mapping. Thus by the Contraction Mapping Theorem we can conclude that Kepler's equation has a unique solution x^*, which can be found by computing the sequence of approximations $\{x_n\}$ via the rule

$$x_{n+1} = a \sin x_n + b,$$

starting with an arbitrary initial approximation x_0.

To see how the shrinking condition arises in a different concrete situation, return to the set of linear algebraic equations (§) given earlier in example 1. Recall that this set of equations can be abstractly written as $Ax = b$. We want to see under what conditions the linear operator (matrix) A is a contraction mapping of the space R^n to itself. This will provide conditions under which the equations (§) have a unique solution for any element b in R^n. A bit of algebra, the details of which can be found in the chapter references, leads to the result that if the inequality

$$\sum_{j=1}^{n} |a_{ij}| < 1,$$

holds for all i, then the set of linear algebraic equations (§) has a unique solution for arbitrary b in R^n. In other words, if the sums of the magnitudes of the elements in every row of A are less than 1, there is a unique solution to the equations (§).

Unfortunately, this condition on the elements of A is pretty restrictive—very restrictive, in fact. This is to be expected, however, because the Contraction Mapping Theorem makes no use of the linear structure of the matrix A; it is a very general result, so we can't expect too much by way of precision in its application. It turns out that the above condition on the elements of A can be greatly weakened by exploiting the linear structure of A. Let's take a few pages now to show just how detailed are the results we can obtain when A is linear.

The Contraction Mapping Theorem ensures the existence of a fixed point in X of the operator A. Because a point x in X can be regarded as a vector, let's consider all the points that lie in the same "direction" as x. Arguing by analogy with points in R^n, which have a direction determined by the line from the origin to the point, we can say that the points in the same direction as x are those of the form λx, where λ is any real number. This set forms a subspace in X, the one-dimensional vector space generated by the element x.

Now consider the directions that are left unchanged by application of the linear transformation A. Any such subspace consists of points x in X for which there is a real number λ such that $Ax = \lambda x$. Values of λ for which this holds are called the *characteristic values,* or *eigenvalues,* of the operator A. Analogously, if λ is such a characteristic value of A, the point x for which $Ax = \lambda x$ constitutes the characteristic vector associated with λ. It turns out that the entire structure of a linear operator A can be read off by knowing its set of characteristic values λ (often termed the "spectrum") and associated characteristic vectors. The situation is especially simple in the case of a finite-dimensional space such as $X = R^n$, but gets rather tricky when X is infinite-dimensional. So let's first hear the finite-dimensional story, and then see some of its generalizations to the infinite-dimensional case.

One of the most basic results in number theory is the so-called Fundamental Theorem of Arithmetic, which states that every integer can be written as a unique product of prime numbers. These primes then constitute the "atoms" from which the entire set of integers can be constructed. The analogous result for linear operators is called the Spectral Theorem, which gives the atoms from which every linear transformation is composed. The simplest version of this result is for the case of a finite-dimensional complex space, $X = C^n$, where for technical reasons it's convenient to allow multiplication of elements by complex numbers rather than the real numbers considered thus far. Here is the result:

SPECTRAL THEOREM (FINITE-DIMENSIONAL CASE) *Assume that* A *is a symmetric linear transformation from* C^n *to itself. Then there exists* *a finite number of distinct real numbers $\lambda_1, \lambda_2, \ldots, \lambda_p$ and of nonempty closed subspaces $M_1, M_2, \ldots, M_p$ of C^n, each number λ and subspace M determined uniquely by A, such that*

1. *The elements of M_i are eigenvectors corresponding to λ_i.*
2. *If $i \neq j$, every element in M_i is orthogonal to the elements of M_j.*
3. *If x is in C^n, there exists a unique element in M_i such that*

$$x = x_1 + x_2 + \cdots + x_p.$$

In other words, each element of C^n can be written as a unique linear combination of elements from the subspaces $\{M_k\}$. ■

As just indicated, the last part of the Spectral Theorem tells us is that the elements of the subspaces corresponding to the eigenvectors of A constitute the atoms from which the transformation A can be built. Moreover, for each point in the space C^n these atoms are unique. The orthogonality condition in part 2 of the theorem is a consequence of the fact that every finite-dimensional space such as C^n is also a Hilbert space, so it makes sense to talk of orthogonality because there is an inner product structure on the space.

Every finite-dimensional space is a Hilbert space, and the Spectral Theorem gives a complete description of the structure of a linear operator on such a space. It would take a bit more technical jargon than we want to go into in this book to explicitly state the generalization of the Spectral Theorem to the infinite-dimensional setting. Suffice it to say that by imposing some reasonably tight technical conditions on the operator A, *exactly* the same result as stated above can be proved when X is a Hilbert space. In fact, essentially the same result can be established even under weaker conditions on the operator A. But these are matters best taken up in the chapter references. For our needs here, what's important is that the linear structure of the operator can be used to give a complete picture of how it is formed and how it acts on a space X.

Happily, though, it's not beyond our brief in this chapter to mention what is certainly the most famous unsolved problem in operator theory, the so-called *Invariant Subspace Problem*. The problem concerns bounded operators on an infinite-dimensional complex Hilbert space X. A subspace M of a Hilbert space is said to be *invariant* under the operator A if Ax is in M for every x in M. There are two trivial invariant subspaces: the zero element in X and the entire space X itself. This fact leads to the Invariant Subspace Problem: *Does every bounded operator on a Hilbert space X have a nontrivial invariant subspace?*

Mathematicians hope that knowledge of the invariant subspaces of operators will illuminate their structure. Even the simplified version of the Spectral Theorem (finite-dimensional case) shows that the answer to this problem is yes, when X is a finite-dimensional Hilbert space. In this case, the Spectral Theorem shows that operators on finite-dimensional spaces are direct sums of particularly well-behaved operators on eigenspaces. So to prove that every operator has a nontrivial invariant subspace might be the start of a general structure theory. On the other hand, a counterexample would lead to some extremely interesting approximation results. If A has no nontrivial invariant subspace, then linear combinations of the elements $x, Ax, A^2x, A^3x, \ldots$ are dense in X for any element x in X. This means that every element x^* in X can be approximated as closely as we wish by a combination of such powers $\{A^n x\}$ for some sufficiently large—but finite—integer n.

Now let's turn our attention back to solutions of equations, and show how the Spectral Theorem can be used to produce the solution to a classic problem in the theory of thermodynamics.

- *Heat Conduction in a Rod:* Consider the problem of finding the distribution of heat on a metal bar of length L whose ends are insulated. Suppose the function $u(x, t)$ describes the heat at a point x on the bar at time t, and that the initial heat distribution at time $t = 0$ is $u(x, 0) = \phi(x)$. Then standard arguments from classical physics show that the evolution of the heat distribution over time is given by the heat equation:

$$\frac{\partial u}{\partial t} = \kappa \frac{\partial^2 u}{\partial x^2},$$

with the boundary conditions $u(0, t) = u(L, t) = 0$ and the initial condition $u(x, 0) = \phi(x)$. This is a linear partial differential equation for the unknown function u in the two variables x and t. Here the parameter κ just measures the heat conduction properties of the metal that constitutes the bar, κ having a large value for a bar of highly conductive copper and a small value for a bar of something like ceramic, which has low conductivity.

The traditional method for solving the heat equation is to make use of the linearity to separate the variables into functions $X(x)$ and $T(t)$, and express the unknown heat distribution function $u(x, t)$ as a product of these two functions, thus $u(x, t) = X(x)T(t)$. Substituting this representation into the partial differential equation leads (after some algebra) to the equations

$$\frac{d^2 X}{dx^2} = -\omega^2 X,$$

$$\frac{dT}{dt} = -\omega^2 \kappa T,$$

for some constant ω. These equations for the functions X and T involve the differential operators d^2/dx^2 and d/dt. Thus, the product of the eigenfunctions of these two operators should produce the solution to the heat equation. It turns out that these eigenfunctions have the specific forms

$$X_n(x) = \sin \frac{n\pi}{L} x,$$

$$T_n(t) = e^{-(\kappa n\pi/L)^2 t}, \quad n = 1, 2, \ldots .$$

Thus, the Spectral Theorem assures us that the function $u(x, t)$ can be expressed as the sum

$$u(x, t) = \sum_{n=1}^{\infty} A_n T_n(t) X_n(x),$$

where the coefficients A_n turn out to be given by the expression

$$A_n = \frac{2}{L} \int_0^L \phi(x) \sin \frac{n\pi}{L} x \, dx, \quad n = 1, 2, \ldots .$$

These operations have been purely formal, and it requires a bit of work to prove that this series does indeed converge to a function that satisfies the original heat equation. By using the Spectral Theorem to decompose the differential operators involved in separating the space and time variables, however, we can quickly attack the problem and decompose it into its atomic constituents.

Our focus in this section is on solutions of equations, so it's natural to wonder how the Spectral Theorem helps advance this cause. To answer this query, we need to return to the notion of duality introduced earlier. Recall that for a finite-dimensional space such as $X = R^n$, we can think of an element x as being a column of n real numbers, and the space dual to R^n can be regarded as having elements x^* that are expressed as *rows* of n numbers. So the dual space is formed simply by interchanging columns and rows. The same trick works for a linear operator A defined on a finite-dimensional space. As was shown earlier, every such operator can be explicitly represented by an $n \times n$ matrix of real numbers. So to form the *adjoint* operator A^*, which acts on the dual space X^*, we simply interchange the rows and columns of A. In the language of matrix theory, A^* is called the *transpose* matrix of A. An operator is *self-adjoint* if $(A^*)^* = A$. Now we can give conditions under which the equation $Ax = b$ has a solution.

Consider the two equations in the spaces X and X^*:

$$Ax = b, \qquad (*)$$
$$x^* A^* = 0 \qquad (**)$$

The following result, termed the Fredholm Alternative Theorem, gives conditions under which equation $(*)$ has a solution for any right-hand sides b in X.

FREDHOLM ALTERNATIVE THEOREM   *A necessary and sufficient condition for the equation $(*)$ to have a solution for a fixed right-hand side* b, *is that* b *is orthogonal to every solution of the homogeneous adjoint equation $(**)$. In particular, if the homogeneous adjoint equation has only the trivial solution* $x^* = 0$, *then the inhomogeneous equation $(*)$ has a unique solution for every* b. ∎

The label "alternative" in the name of this theorem comes from the fact that the theorem says that *either* the homogeneous equation $(**)$ has a nontrivial solution *or* the inhomogeneous equation $(*)$ has a unique solution—but not both. Let's conclude this section on linear operators and their structure by showing how the ideas given here provide the mathematical muscle to describe the world of atoms and molecules. We do this by looking at one of the great application areas for functional analysis, the theory of quantum phenomena.

Quantum Mechanics and Functional Analysis

In the 1920s, while functional analysis was being developed in mathematical centers in Germany, Poland, Austria, and elsewhere in Europe, physicists such as Erwin Schrödinger, Werner Heisenberg, Niels Bohr, Camille Pascual, and Max Born were busy laying the foundations for a new physical theory of matter, the quantum theory. The special genius who had a foot in both camps was John von Neumann, who saw immediately how the ideas and concepts of functional analysis could be used to provide a coherent mathematical foundation for quantum mechanics. Here, in slightly more modern form, are the basic ingredients of what von Neumann had in mind.

First and foremost, a quantum system such as an electron is described by a complex Hilbert space X and a collection of self-adjoint linear operators acting on X. There are several components to this description:

1. *Physical states:* The points ψ of X constitute the states of the quantum system. For mathematical reasons, it's especially convenient to think of these states as being the unit vectors in X. That is, the states are such that

$$\|\psi\|^2 = \langle \psi, \psi \rangle = 1.$$

   Intuitively, each physical state of the system corresponds to a mathematical state ψ. Modern developments since the original work in the 1920s have modified this a bit. But for the rough description given here, it's safe to just say that every unit vector ψ of X corresponds to a physical state of the system.

2. *Observables:* Self-adjoint operators on X are called *observables* of the quantum system. Again, intuition says that there is such an operator for each physical "quantity"—position, momentum, spin—associated with the system. Of special importance is the system's energy, which corresponds to an operator H. The jargon term for the energy operator is the *Hamiltonian* of the system.

3. *Measurements:* Let the system be in the state ψ, and suppose we measure the observable A. In contrast to classical physics, for which such a measurement would produce a definite result, in quantum theory the prediction of a measurement's outcome is

only statistical. So we must describe the result of the measurement of A using statistical terms. In this direction, the number

$$\mathcal{E} = \langle \psi, A\psi \rangle,$$

corresponds to the *expected value* of the observable A in the state ψ. By the same token, the number

$$(\Delta \mathcal{E})^2 = \langle A\psi - \mathcal{E}\psi, A\psi - \mathcal{E}\psi \rangle = \| A\psi - \mathcal{E}\psi \|^2,$$

is the *variance* of the observable A in the state ψ.

4. *Dynamics:* The *Schrödinger equation*

$$i\hbar \frac{d\psi}{dt} = H\psi(t), \quad \psi(0) = \psi_0,$$

describes the time evolution of the quantum system. Here ψ_0 is the state of the system at time $t = 0$ and $\hbar = h/2\pi$, where h is Planck's quantum of action. Often, the function $\psi(t)$ is called the *wave function* of the system. We can formally integrate the Schrödinger equation, obtaining the expression for the wave function at time t as

$$\psi(t) = e^{-itH/\hbar}\psi_0,$$

where H is the Hamiltonian for the system.

These are just definitions. Let's justify them by showing how the mathematical objects involved in the definitions can be used to physically interpret the properties of a quantum system.

- *Eigenvalues:* Suppose λ is an eigenvalue of the observable A with corresponding eigenstate ϕ. In other words,

$$A\phi = \lambda\phi, \quad \langle \phi, \phi \rangle = 1.$$

Then

$$\mathcal{E} = \langle \phi, A\phi \rangle = \langle \phi, \lambda\phi \rangle = \lambda\langle \phi, \phi \rangle = \lambda,$$
$$\Delta A = \| A\phi - \mathcal{E}\phi \| = \| A\phi - \lambda\phi \| = \| A\phi - A\phi \| = 0.$$

This little calculation says that when the system happens to be in an eigenstate of the observable A, measuring the system in that state produces the "sharp" value λ. Thus, the possible "pure" values that can be seen when measuring the observable A coincide with the eigenvalues of the operator representing the observable A. But to happen to find the system in an eigenstate of the observable A when we carry out a measurement seems rather unlikely. How unlikely? Our next result answers this question.

• *Probability of an Eigenstate:* Let $\{\phi_n\}$ be an orthonormal set of eigenstates of the observable A, with associated eigenvalues $\{\lambda_n\}$. Thus,

$$A\phi_n = \lambda_n\phi_n, \qquad \text{with} \qquad \langle\phi_j, \phi_k\rangle = \begin{cases} 1, & j = k, \\ 0, & j \neq k. \end{cases}$$

Now, suppose the quantum system is in the state ψ at time t. We would like to know the probability that ψ is one of the eigenstates ϕ_n. To compute this likelihood, we use the fact that every state ψ can be written as an expansion of the eigenstates as follows:

$$\psi = \sum_{n=1}^{\infty}\langle\phi_n, \psi\rangle\phi_n.$$

But since $\langle\psi, \psi\rangle = 1$, we have

$$1 = \langle\psi, \psi\rangle = \sum_{n=1}^{\infty}|\langle\phi_n, \psi\rangle|^2.$$

Therefore, we can interpret the quantity $|\langle\phi_n, \psi\rangle|^2$ as being the probability that the system is actually in the eigenstate ϕ_n. (*Note:* These quantities are usually called the Fourier coefficients of the state ψ.)

Using these probabilities, we can now compute the expected value of the measurement of the observable A and its variance in terms of the pure values (eigenvalues of A):

$$\mathcal{E} = \langle\psi, A\psi\rangle = \sum_n \lambda_n|\langle\phi_n, \psi\rangle|^2,$$

$$(\Delta\mathcal{E})^2 = \|A\psi - \mathcal{E}\psi\|^2 = \sum_n (\lambda_n - \mathcal{E})^2|\langle\phi_n, \psi\rangle|^2.$$

Now let's turn to what is surely the most well-known result of the quantum theory, the celebrated Heisenberg Uncertainty Principle. In a Hilbert space setting, this follows directly from the Schwarz inequality

$$|\langle x, y\rangle| \leq \|x\|\|y\|,$$

for any two elements x and y of a Hilbert space X.

- *Uncertainty Principle:* Suppose that A and B are two observables on a Hilbert space X, and ψ in X is a state of the quantum system. Let $\mathcal{A}$ and $\mathcal{B}$ represent the expected values of the observables A and B, respectively. Then the square roots of the variances of A and B satisfy the relation

$$\Delta\mathcal{A}\Delta\mathcal{B} \geq \mathcal{C},$$

where

$$\mathcal{C} = \tfrac{1}{2}\langle (BA - AB)\psi, \psi \rangle.$$

(*Note:* $\mathcal{C}$ is just the expected value of the compound observable $BA - AB$.)

In rough terms, this relation says that if the two observables A and B do not commute (that is, $AB \neq BA$), then it is impossible to measure the two corresponding physical quantities without error at the same time. For example, if A is the position of a particle and B is its momentum, A and B do not commute; hence, position and momentum cannot be simultaneously measured to arbitrary accuracy. On the other hand, if B were energy rather than momentum it would be possible to measure the two observables as accurately as we wish because the observables position and energy do commute.

To prove the Uncertainty Principle, we first note that $\mathcal{A} = \langle \psi, A\psi \rangle$ and $\mathcal{B} = \langle \psi, B\psi \rangle$. Thus, the Schwarz inequality gives

$$\Delta\mathcal{A}\Delta\mathcal{B} = \|(A - I\mathcal{A})\psi\|\|(B - I\mathcal{B})\psi\| \geq |\langle (A - I\mathcal{A})\psi, (B - I\mathcal{B})\psi \rangle|$$
$$\geq |\Im\langle (B - I\mathcal{B})(A - I\mathcal{A})\psi, \psi \rangle| \geq \tfrac{1}{2}|\langle (BA - AB)\psi, \psi \rangle|.$$
$$(\sharp)$$

(*Note:* In the expression ($\sharp$), $\Im$ denotes the imaginary part of the inner product $\langle (B - I\mathcal{B})(A - I\mathcal{A})\psi, \psi \rangle$, and I denotes the identity operator, whose action is to produce the same element upon which it operates.)

To this point, most of our discussion has been centered on linear operators in a Hilbert space because this is the setting in which we can be most explicit about the structure of the operators and what they do. But to speak of linearity in mathematics is like giving an account of a trip to the zoo by talking about all the elephants. They may be the biggest objects in the zoo, but most of what's on offer in mathematics—and in the zoo—is not elephants. So let's conclude our account of functional analysis with a brief look at the wild and wooly world of nonlinear operators.

The Importance of Being Nonlinear

Consider the space of points in the plane, R^2. If f is a linear functional defined on these points, then the set of pairs $x = (x_1, x_2)$ satisfying $f(x) = c$, where c is a fixed constant, is geometrically a straight line. To be even more specific, we know by the Riesz Representation Theorem that every linear functional on R^2 has the form $f(x) = \langle x, y \rangle$ for some $y = (y_1, y_2)$ in R^2. So there is no loss of generality in writing the equation $f(x) = c$ as $y_1 x_1 + y_2 x_2 = c$. This is the equation of a straight line in the (x_1, x_2) plane. For this reason, linear functionals and operators give rise to geometrically "flat" objects. But what about nonlinear operators? What kinds of geometrical objects do they generate?

To address this question, consider the simple quadratic functional g on R^2 given explicitly by $g(x) = g(x_1, x_2) = ax_1^2 + bx_2^2$. If we let k be a fixed constant, then the set of points x in R^2 satisfying $g(x) = k^2$ is a curve, not a straight line. In fact, it is a circle. This suggests that nonlinear functionals give rise to curved hyperspaces in exactly the same way that linear functionals generate flat spaces. This indeed turns out to be the case. So linearity is to flat geometries as nonlinearity is to curved ones.

Our focus in this chapter is on solving equations in infinite-dimensional spaces. So we confine our discussion of nonlinear functionals to this end, in particular to providing conditions under which the equation $G(x) = 0$ has a solution in an abstract space X. Our first general result in this direction is the extension of the classical Brouwer Fixed-Point Theorem discussed in passing in Chapter 2 of *Five Golden Rules* to the case when X is a Banach space. In this setting, the result is known as Schauder's Fixed-Point Theorem; its formal statement follows.

SCHAUDER'S FIXED-POINT THEOREM *Let* M *be a bounded, closed, convex nonempty subset of a Banach space* X. *Suppose further that* A *is a continuous transformation of* M *to itself, such that for every bounded sequence* {x_n} *of elements of* X, *there exists a subsequence* {x'_n} *for which the sequence* {Ax'_n} *converges to an element of* M. *Then there exists a point* x* *in* X, *such that* Ax* = x*, *that is,* A *has a fixed point in* M. ■

(*Note:* The reader should recall from Chapter 2 of *Five Golden Rules* that a set is convex if the straight line joining any two points of the set remains entirely within the set, and that a set is closed if it contains all its boundary points.) The conditions imposed on the operator A in the statement of Schauder's Theorem define A to be what is called a *compact* operator on M.

Schauder's result shows that there is some solution to the equation $Ax = x$ in the subset M of the Banach space X. Here is a closely related result that illustrates a very important principle in mathematics, namely, that if you can impose a priori estimates on the magnitude of a putative solution to the equation $Ax = x$, then you can prove the existence of such a solution. The result is known as the Leray-Schauder Principle.

LERAY-SCHAUDER PRINCIPLE   *Suppose a compact operator* A *maps the Banach space* X *to itself, and that there is a number* r > 0 *such that if* x* *is a solution of the equation* Ax $=$ x, *then* $\|x^*\| \leq$ r. *Then the equation* Ax $=$ x *has a solution.*   ∎

It should be noted here that we do *not* assume that the equation has a solution; rather, we say that if there is a solution, then that solution is uniformly bounded in the norm. This is the a priori condition, and is trivially satisfied if there is a number $r > 0$ such that $\|Ax\| \leq r$ for all x in X. Thus follows the principle that "a priori estimates yield existence."

One of the more interesting applications of this result lies in the area of fluid mechanics, where the Leray-Schauder Principle is used to prove the existence of a solution to the famed Navier-Stokes equations. These equations describe the pressure and velocity of a fluid moving in a fixed region, and form the basis for the greatest mystery of classical physics: turbulence. Although the full details are much too complicated for presentation here, we should at least state the problem in formal terms.

Let G be an open, connected, bounded region in three-dimensional space R^3. The steady-state motion of an incompressible viscous fluid in

G is governed by the Navier-Stokes equations, which take the form:

$$\eta\Delta\mathbf{v} + \rho(\mathbf{v}\Delta)\mathbf{v} = \mathbf{K} - \nabla p, \quad \text{on } G \quad \text{(equation of motion)},$$
$$\Delta\mathbf{v} = 0, \quad \text{on } G \quad \text{(incompressibility conditions)},$$
$$\mathbf{v} = 0, \quad \text{on } \partial G \quad \text{(boundary condition)}.$$

In the foregoing equations, $\mathbf{v}(x)$ is the velocity vector of the fluid at the point x in G, $p(x)$ is the pressure at x, ρ is the fluid density, $\mathbf{K}(x)$ is the external force density at x, and $\eta > 0$ is the fluid's viscosity. Moreover, the symbols Δ and ∇ are the Laplacian, $\Delta f = \partial^2 f/\partial x^2 + \partial^2 f/\partial y^2 + \partial^2 f/\partial z^2$, and the gradient operator, $\nabla f = (\partial f/\partial x, \partial f/\partial y, \partial f/\partial z)$, respectively. Physical experiments show that turbulence occurs if the external forces $\mathbf{K}$ become too large, a fact that complicates greatly the mathematical analysis of when, exactly, the Navier-Stokes equation has a solution $(\mathbf{v}, p)$ satisfying the above equation and boundary conditions.

The basic approach to proving existence of such a pair of functions $(\mathbf{v}, p)$ is to first *assume* there exists such a solution, and then show it satisfies the equation and boundary conditions. This assumption generates some *necessary* conditions that any solution must satisfy. These conditions take the form of orthogonality relations, strongly suggesting that the right mathematical setting for the problem is in a Hilbert space, whose elements are functions having certain properties. These spaces are now termed *Sobolev spaces,* after the Russian mathematician S. Sobolev, who first explored their properties. Next, one assumes a velocity field $\mathbf{v}$ that satisfies the equation plus the necessary conditions, and then seeks conditions on the pressure function $p(x)$ that also satisfies the Navier-Stokes equation. This procedure gives rise to *sufficient* conditions for solvability of the pressure equation. The reader interested in following the details of this procedure is invited to consult the references cited for this chapter.

As with the famed Brouwer Fixed-Point Theorem considered in *Five Golden Rules,* Schauder's generalization to the Banach space setting offers no procedure by which to actually construct the solution to the equation $Ax = x$; it is simply an existence result, saying that there is a point in X satisfying the equation. Interestingly, to find the asserted solution, we turn to a method originally developed by Newton nearly 300 years ago.

To motivate the overall idea of Newton's method, consider the scalar equation

$$F(x) = x^2 - a = 0, \quad a > 0.$$

The solution of this equation is the square root of the positive real number a, or, equivalently, the root of the equation $F(x) = 0$.

The basic technique of the method is illustrated in Figure 4.8. At a given point, the graph of the function F is approximated by its tangent, and an approximate solution to the equation $F(x) = 0$ is taken to be the point where the tangent crosses the x axis. The process is then repeated from this new point. This procedure defines a sequence of points according to the rule

$$x_{n+1} = x_n - \frac{F(x_n)}{F'(x_n)}, \tag{b}$$

where F' denotes the first derivative of the function F. In the particular case above, where $F(x) = x^2 - a$, Newton's method gives the very rapidly converging scheme for calculating the square root of a,

$$x_{n+1} = \tfrac{1}{2}\left(x_n + \frac{a}{x_n} \right).$$

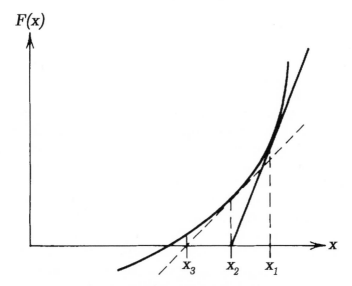

Figure 4.8. Newton's method.

Now, suppose that instead of a scalar equation in the space $X = R^1$, we have F being a nonlinear operator between two Banach spaces X and Y, and we want to solve the equation $F(x) = 0$. Proceeding entirely formally, the iterative rule (♭) above becomes

$$x_{n+1} = x_n - \left[F'(x_n)\right]^{-1} F(x_n), \quad x_0 = \bar{x}.$$

In this expression, we have to specify what we mean by the term F' because F is now an operator, not just a simple function. We'll come to this in a moment. For now, let's observe that the iterative rule (♭) can be viewed as an application of the method of successive approximation applied to the operator $P(x) = x - [F'(x)]^{-1} F(x)$. The connection to fixed-point theory is now clear: a fixed point of P is a solution of $F(x) = 0$.

The question of how to interpret the derivative F' of the operator F would take us too far afield to pursue here. Let's just say that functional analysts have sorted out this question, and the answer is to use what is called the *Fréchet derivative* as the operator-theoretic analogue of the ordinary derivative that we know from calculus. What we need to know now is under what circumstances this iterative procedure actually converges to a solution of the equation $F(x) = 0$.

Roughly speaking, what can be shown is that Newton's method converges provided the initial approximation $\bar{x}$ is sufficiently close to the solution x^*. What is especially attractive about Newton's method is the very rapid rate of convergence, a nice reward for the possibly difficult job of finding the inverse of the operator $F'(x_n)$ at each step of the process. Detailed analysis shows that under mild restrictions on the nature of the operator F, the method converges *quadratically* provided the initial approximation is reasonably close to the actual solution. This means that if the solution x^* is a numerical quantity, such as the square root operation shown earlier, the number of correct digits approximately doubles with each successive iteration. To illustrate this point, Table 4.2 shows the use of Newton's method to calculate the square root of $a = 10$, starting with the initial approximation $\bar{x} = 3.0$.

Considerably more could be said about the generalization of results for nonlinear operators on finite-dimensional spaces to the infinite-dimensional setting. But space precludes their discussion here, so the curious reader will have to consult the chapter references for discussions at all levels of detail. What we can say in concluding this brief

Table 4.2. Newton's method for computing $\sqrt{10}$.

Iteration	x_n	x_n^2
1	3.0000000000	9.0000000000
2	3.1666666667	10.0277777778
3	3.1622807018	10.0000192367
4	3.1622776602	10.0000000000

excursion into mathematical abstraction and generalization is that abstraction is what functional analysis is all about. The entire field is devoted to providing a single framework within which we can unify calculus, fixed-point theory, singularity theory, and much, much more. In this sense, functional analysis plays the same role in mathematics that the theory of elementary particles plays in physics or that the periodic table of the elements plays in chemistry. It serves as a kind of organizing center around which to encompass a great deal of mathematical analysis beneath a single umbrella. Who could ask for more?

CHAPTER

5

The Shannon Coding Theorem

Information Theory

Communication, Information, and Life

In everyday life, people use the word "information" in many different ways. One of the most common is as a measure of novelty or surprise. So, for example, if my neighbor tells me that he went to work yesterday, that doesn't tell me much because he goes to work every day. On the other hand, if he says that he's decided to divorce his wife and move in with his secretary, that's really surprising news because I always thought his marriage was on solid ground. So the first case contains little, if any, information because he's not telling me anything that I didn't already know, or at least strongly suspect. The news that my neighbor is running off with his secretary, however, is laden with information because it's something completely unexpected to me, hence a priori a low-likelihood event.

Information theory as we know it today had its origins with problems of electrical communication of signals. In particular, in the 1920s, engineers H. Nyquist and R. V. Hartley, working at the Bell Telephone Laboratories, developed methods for measuring the information content in a string of symbols transmitted through a communication line. Then, during World War II, it became important to predict the course of airplanes from inaccurate or "noisy" radar data. This raised the important question: Suppose we have a varying electric current that represents data concerning the present position of an airplane, and meaningless additional current is added to this signal. How can we best filter the total received current, consisting of the signal plus the noise, so as to make the best possible estimate of the signal? This problem was solved in Russia by the great mathematician Andrei N. Kolmogorov and in the United States by MIT mathematician Norbert Wiener. These studies in the first half of the twentieth century set the stage for the formal development of what we now call information theory, which in many ways is the brainchild of Claude Shannon, a communications engineer at Bell Telephone Laboratories.

In 1948, Shannon created a mathematical theory for measuring information such as the noise problem above. Because we don't know in advance what combination of symbols might be transmitted through the communication channel, we can only attach a probability to the transmission of any particular symbol. Shannon's measure, then, was a very simple one. If the a priori probability of an event E (a particular symbol being transmitted) is $P(E)$, then the *information* content $I(E)$ of the event E actually occurring is simply the negative of the logarithm of the probability of that symbol being sent. Mathematically,

$$I(E) = -\log P(E),$$

where log is the logarithm function. The base used in calculating the logarithm determines the unit of information measure. The most common unit is to use base 2 for the logarithm, in which case the information is measured in binary digits, or bits. This is because the binary number system is especially convenient both for computers and for proving theorems, because it contains only the two elements, 0 and 1. Moreover, there is no loss in mathematical generality in using base 2 over using any other base such as 10, 8, or 16. If, for instance, we used the more familiar base 10, then the information content would be measured in decimal digits, whereas the use of base 8 would presumably lead to something called an "octit." So we use this binary measure throughout the remainder of this chapter. Thus, in what follows log should be interpreted as $\log_2$, and all information is then measured in bits. For example, then, if you flip a fair coin and it turns up heads, the binary information content in such a result is

$$
\begin{aligned}
\text{Information content of heads } &= I(\text{heads}) \\
&= -\log P(\text{heads}) \\
&= -\log \tfrac{1}{2} \\
&= -(\log 1 - \log 2) \\
&= -(0 - 1) \\
&= 1 \text{ bit}
\end{aligned}
$$

Shannon's measure of information is just that: a measure—a proposal if you like, not a definition. We could think about measuring information as some other function of the likelihood of an event beside the logarithm of the probability of the event, subject only to the constraint that the information content of the event be larger when the event's likelihood is smaller. For example, the reciprocal measure $I(E) = 1/P(E)$ is also a possibility. However, it's easy to see that this measure is not additive, in the sense that the information in two separate events does not equal the sum of the information in the individual events, as it does with the logarithmic measure used by Shannon. But additivity is important because if the events are totally independent and thus do not interfere with each other, it stands to reason that the information content of the two events together is just the sum of their individual information levels.

For example, suppose we flip a fair coin and at the same time roll a single die. Now consider the two events heads on the coin and a six on the die. There is no reason to believe that either of these events influences the occurrence of the other, so the information content of the compound event $E = \{\text{heads}\} + \{6\}$ is simply $I(\text{heads}) + I(6)$. So the main attraction of Shannon's measure is that it accords with our intuition—and it has some nice mathematical features that make it convenient to use.

As an aside, it's amusing to see that the logarithmic measure of information works pretty well for the concept of information as we humans use it because it essentially measures the "surprise" we experience from the occurrence of an event. But not all organisms transmit information along these lines. For example, there is evidence to show that ants transmit locational information about food to each other in a time proportional to the actual number of different possible locations where the food resides rather than in a time proportional to the logarithm of the number of locations; the linear function ants use always dominates the logarithm function. Another example is the way the phone company has set up its multiplexing schemes, so as to exploit the optimality of the logarithm function for moving information from one place to another. This means that the ants are not transmitting the information as efficiently as they could. That is, their information transfer rate is slower than it might be. To support this claim, we need a few more notions about the information contained in a system, such as the states of a financial market or the signals sent by a radio transmitter, and the way it is encoded into a set of standard symbols such as 0 and 1 or a, b, c for transmission.

Ludwig Boltzmann, a Viennese physicist of the late nineteenth century, is perhaps best remembered for his work in thermodynamics and the connections he discovered between the theory of heat and the more general issues of randomness and order. He is today credited with having introduced the notion of *entropy* as a measure of the disorder present in a collection of objects of any sort—an idea we have already seen has served as the basis for the theory of information, which turned out to be so crucial to the development of modern communications technology. In fact the formula $S = k \log W$, expressing the entropy S as being proportional to the logarithm of W, the number of possible states that a system can assume, is engraved on Boltzmann's tombstone in Vienna's Zentralfriedhof, a fitting memorial to the importance of this fundamental idea. In this expression, the constant of proportionality k is even today termed *Boltzmann's constant* in recognition of this magnificent achievement. But at the time he was carrying out this pioneering work, the achievement was anything but magnificent, at least if one was listening to the leading scientists of the day.

Boltzmann's problem was that his theory of heat involved an assemblage of atoms moving according to the usual rules of Newtonian mechanics. He used this concept of an atom as a particle of matter to construct his theory of heat as a statistical property emerging out of the overall motion of these atoms. This idea was put forth around the turn of the century, several years before the work of Ernest Rutherford, J. J. Thomson, and Niels Bohr gave the concept of an atom its modern birth. As a result of his atomistic speculations, Boltzmann came into heated conflict with several of the giants of the scientific community, most notably his Viennese colleague Ernst Mach and the German physical chemist Wilhelm Ostwald, who argued forcefully against the idea of the atom. Ostwald, in particular, preferred a theory of heat based upon the notion of energy rather than matter. Depressed by the acrimony of this opposition, as well as his failing eyesight and what he thought of as the decline of his mental faculties, Boltzmann took his life in Duino, Italy, on September 5, 1906. But let's have a harder look at how Boltzmann's vision has been translated into action by modern communication theorists.

The average information contained in a system is called its *entropy*. This is a measure of the uncertainty in a system at a given moment because the more information in a system, the greater the uncertainty in specifying exactly in what state the system is. So if the system S can be in one of n possible states, $s_1, s_2, \ldots, s_n$, and the probability of it being in state s_i is $P(s_i)$, then the entropy of the state s_i is simply $P(s_i) \log P(s_i)$. Because the states are independent, in the sense that the system S can be in only one of them at any given time, the total entropy of S, $H(S)$, is simply the sum of the entropies of each of the possible states, or, more formally,

$$H(S) = -\sum_{k=1}^{n} P(s_k) \log P(s_k).$$

Sometimes this measure of average information is called the *negentropy* because it actually is the negative of the entropy as defined by Boltzmann in his work on thermodynamics.

It is an easy exercise in calculus to see that the entropy $H(S)$ reaches its maximum if each of the n states is equally likely (occurs with probability $1/n$), and is at its minimum if all of the states except one cannot possibly occur. In this case, $H(S) = 0$. These extreme cases give a good intuitive account of what entropy means: it is a measure of the surprise, or novelty, experienced when the system is actually seen to be in one of its states. So if beforehand we believe any of the n states to be equally likely as the outcome of an observation of the system, then the surprise at seeing one of the states turn up is high; on the other hand, if before the observation we *know* the system is in a particular state, there is no surprise at all when we carry out the observation and confirm that it is indeed in that state. In short, the observation in this case produces no information. So, in general, the entropy represents the information that can be obtained by actually observing the system's state.

As an example, suppose we toss a loaded die that's weighted so that the faces $1, 2, \ldots, 6$ occur with likelihoods 1/6, 1/6, 1/3, 1/6, 1/6, and 0. Thus, a 6 will never occur, whereas the outcome 3 is twice as likely as any of the other nonzero possibilities. In this case, the entropy of the loaded die is

$$H(\text{die}) = 4 \times \frac{1}{6} \times \log \frac{1}{6} + \frac{1}{6} \log \frac{1}{3} + 0$$

$$= - \left[\frac{2}{3} \log 6 + \frac{1}{6} \log 3 \right]$$

$$= - \left[\frac{2}{3} + \frac{5}{6} \log 3 \right]$$

Note that this number is smaller than $-\log 6 = -(1 + \log 3)$, which is the entropy of an unloaded die.

Information is of little use if it cannot be transmitted from one point in space and/or time to another. For example, the Rosetta stone has passed information to us over the course of time, whereas a television signal transmits mostly over space (although time necessarily enters too, because the speed of the signal is not infinite). This introduces the idea of a communication channel through which information is conveyed from its sender to a receiver. As we are all well aware, communication channels are tricky things, and one of the biggest issues in information theory is how to neutralize the distortion of information introduced by the channel because they almost always seem to corrupt the information passing through them in wild and wondrous ways. A good example is the transmission of a joke or story from one person to another. How many times have you heard a joke one day, then a few weeks later heard the same joke again—but with a completely different lead-in, or even a different punchline? More common examples of this same phenomena are the static on a telephone line and the mutations caused by cosmic rays in a strand of genetic material.

To address the problems of information transmission through a possibly noisy communication channel, Shannon created a mathematical model of the communication process whose structure is shown in Figure 5.1. Here we see a *message source,* which might consist of textual symbols on a computer keyboard or phonemes spoken into a telephone. The message then arrives at a *transmitter,* which changes the message into a signal that is sent through the *communication channel.* The channel usually distorts the signal in various ways before it reaches the *receiver,* whose function is to intercept the signal and reconstruct it into the original message for presentation at the *destination.* In this setup, it's important to note that the processes by which the message comes to the receiver and the way the receiver converts the signal into a message are carried out with no distortion. So the only noise in this system resides in the communication channel itself.

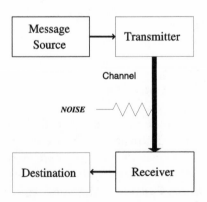

Figure 5.1. Shannon's model of a communication process.

The noise introduced into the channel can be represented by expressing the probabilities that particular messages are transmitted when another message is received. So, for instance, assume the source sends n different messages $m_1, m_2, \ldots, m_n$ over the course of time, while the destination receives the messages $r_1, r_2, \ldots, r_n$. Then the operation of the channel is described by the set of probabilities,

$P(m_i|r_j) = $ probability that message m_i is transmitted when message r_j is received,

for $i, j = 1, 2, \ldots, n$. For ease of writing, we will abbreviate this probability as simply P_{ij}. Thus, a channel with no noise would have $P_{ij} = 0$, for $i \neq j$, and $P_{ii} = 1$.

A quantity of great interest is the actual amount of information transmitted by the channel as a result of the receipt of a message r_j. This is now easy to calculate. Before the message was received, our best estimate of the probability that message m_i was sent is $P(m_i)$. After receiving the message r_j, the probability that the message m_i was sent is, by definition, P_{ij}. The gain in information in seeing message r_j when we have estimated that we would see message m_i is denoted I_{ij}. This is the difference of the information content of the two quantities, $P(m_i)$ and P_{ij}. But because the information content is defined in terms of logarithms, mathematically we have

$$I_{ij} = -\log P(m_i) - [-\log P_{ij}]$$
$$= -\log P(m_i) + \log P_{ij}$$
$$= \log \frac{P_{ij}}{P(m_i)}.$$

A bit of algebraic sleight-of-hand transforms this into the much more transparent expression

$$\log \frac{P(m_i, r_j)}{P(m_i)P(r_j)},$$

where $P(m_i, r_j)$ is the probability that m_i was sent and r_j was received. From this form we immediately see that the information gain is 0 if the messages m_i and r_j have no connection to each other (are statistically independent) because in that case the probability $P_{ij} = P(m_i)P(r_j)$. On the other hand, if the two messages are correlated, then the joint probability will be much larger than the product of their individual probabilities;

hence, the information gain will be positive. This is because a correlation between the two messages means that knowledge of one conveys *some* knowledge about the other.

The foregoing analysis relates to two specific messages, m_i and r_j. But it can be extended to compute the average information gain by using averages over all messages instead of probabilities for specific messages. Recalling that $H(S)$ is the entropy of the entire set of messages, and $H(S|D)$ is the entropy of the message source *given* the received D, this leads to the expression

$$I(S; D) = H(S) - H(S|D)$$
$$= H(S) + H(D) - H(S|D).$$

Here $H(D)$ is the analogous entropy of the receiver (destination), and $H(S|D)$ is the average information content in the statement "m_i was sent given that r_j is received." Formally,

$$H(S|D) = -\sum_i \sum_j P(m_i, r_j) \log P_{ij},$$

where $P(m_i, r_j)$ is the probability defined previously.

Now, how much information can a given channel actually carry? This leads us to consider the notion of channel capacity.

The *capacity* of the communication channel is the maximum information that can be sent between the source and the destination. Note that the entropy of the destination, $H(D)$, plays no role in the channel capacity because the information has already passed through the channel by the time it reaches the destination. Thus, the information actually received at the destination is the information in the source ($H(S)$) minus the reduction in information by noise in the channel ($H(S|D)$). Mathematically, this is simply

$$C = H(S) - H(S|D),$$

The following example of color discrimination uses these same concepts.

• *Example—Color Detection:* Suppose a subject is asked to distinguish between two colors, say red (R) and green (G). So here the two messages are $m_1 = R$, $m_2 = G$. If the two colors are shown to a color-blind person, that person might identify them correctly 75 percent of the time. If so, we then have the probability that green is seen when green is sent (P_{11}) is 0.75. Similarly, for red $P_{22} = 0.75$, and $P_{21} = P_{12} = 0.25$. The capacity of this simple channel is $C = 1 - H(S|D)$, since at most one bit of information can be transmitted through a channel in which only two possible messages may pass (what is called a *binary* channel). Here $H(S|D)$ is the entropy of the source S (which can send either red or green), given that the receiver D sees red or green. Putting this into numbers, the capacity in this color-blind situation turns out to be

$$
\begin{aligned}
H(S|D) = C &= -\left[P(G|G) \log P_{11} + P(R|G) \log P_{12} + \right. \\
&\qquad \left. P(G|R) \log P_{21} + P(R|R) \log P_{22} \right] \\
&= -\left[\frac{3}{8} \log \frac{3}{4} + \frac{1}{8} \log \frac{1}{4} + \frac{1}{8} \log \frac{1}{4} + \frac{3}{8} \log \frac{3}{4} \right] \\
&= -\left[\frac{3}{4}(\log 3 - 2) + \frac{1}{4}(0 - 2) \right] \\
&= 2 - \frac{3}{4} \log 3 = 2 - \frac{3}{4}(1.585) = 2 - 1.188 \\
&= 0.81 \text{ bits.}
\end{aligned}
$$

So if we are faced with a noisy communication channel, we need to find a way to overcome the noise and have the destination receive the same message that was transmitted. The key to solving this problem lies in good schemes to both *encode* the message at the transmitter and *decode* it at the destination. The following story illustrates the situation.

During the American Revolutionary War, British troops were coming to Boston to arrest revolutionaries Samuel Adams (now more well known for his brewing than his politics) and John Hancock, as well as to capture supplies. The colonial leader, Paul Revere, arranged a scheme by which he would signal the direction from which the British would be coming from Boston's Old North Church. The agreed-upon signals, which were to be implemented by lanterns in the church belltower, were as follows:

Code	Meaning
no light	direction unknown
1 light	attack by land
2 lights	attack by sea

This code has since given rise to the famous expression, "One if by land, two if by sea."

In the actual situation, Paul Revere's coding scheme worked fine, as two lanterns were shown—and properly decoded—to indicate an attack from the sea. But information theorists might wonder whether a different coding scheme might have worked better. Could Revere have hit on a method that used fewer lights, on the average? Is there a coding procedure that is less prone to misreading by the colonists looking up at the belltower? Coding theory provides an answer to these questions.

Paul Revere's coding scheme introduces the two most important problems in communication theory addressed by Shannon's and subsequent work.

1. *Noiseless Coding:* What is the minimal amount of information we need in a coding scheme to uniquely decode any message sent over a noise-free channel?

2. *Noisy Channel:* What is the maximum amount of information that can be transmitted via a noisy channel with an arbitrarily low rate of error?

Problem 1 asks about the best way to code the source message to be able to reconstruct the message at the destination if the coded message is sent via a noiseless channel. Shannon's Noiseless Coding Theorem solves this problem. In Problem 2, on the other hand, we are confronted with a channel that distorts messages that pass through it. So we ask if there is some way of coding the source message, so that what appears at the destination can be reconstructed with arbitrarily small error. The Shannon Coding Theorem tells us under what circumstances such an encoding of the message is possible. To get to these two fundamental results of information theory, we need to know a lot more about what constitutes a code.

Making Codes

Virtually all information that is stored or transmitted is expressed in a code of some sort. Natural languages, such as English or Chinese, are certainly the most familiar codes—so familiar, in fact, that we don't usually even think of them as codes. But they are. For example, the English word "water" is simply a code using the symbols of the alphabet for the liquid substance that we drink and use to wash our clothes. Similarly for the German word "Wasser," which codes the very same substance using a different code.

Unlike English and German, however, most codes are of fixed length. This means that the words in the code are formed by "stringing together" the same number of symbols for each word. The famous genetic code hardwired into every cell in your body is a good example of such a code. In this case, the symbols of the code are the four nucleotide bases adenine (A), guanine (G), cytosine (C), and uracil (U). The code words are strings of these symbols, each of length 3. Each one of these "words," or codons, in the language of molecular biology, represents either one of the 20 amino acids making up the proteins of life or a punctuation mark, telling the cellular protein manufacturing machinery that the protein it is constructing has been formed. The message source in this situation is the cellular RNA, which is read, three symbols at a time, to produce the code words. The words are then transmitted through a channel, consisting of the ribosomal genetic translation machinery, to the destination, which is the set of 20 amino acids from which all proteins are formed. The genetic code is shown in Figure 5.2, where, for example, we see that the three-symbol sequence *UCA* corresponds to the amino acid serine, whereas the code word *GAA* represents glutamic acid.

Because there are four possible nucleic acids that can appear at each location in a code word, each position represents $\log 4 = 2$ bits of information. Now because each codon consists of three symbols, each of which represents two bits of information, a codon can carry $2 \times 3 = 6$ bits of information. Because amino acids may be formed by more than one codon (for example, any codon of the form *GC?* represents the amino acid serine), it can be shown that the six bits of information in a codon are reduced to around 4.3 bits. Thus, the rate of information transfer with our earthly genetic coding scheme is 4.3 bits/6 bits = 0.72. We'll come back to this point later when we speak about error correction and noisy channels.

Another type of fixed-length code that just about everyone nowadays has encountered is the ASCII code, by which computers represent alphanumeric characters. ASCII provides 128 code words for 128 characters, each code word being a string of 0s and 1s of length 7. Thus, each such codeword contains seven bits of information. Just for the sake

	U	C	A	G	
U	phenylalanine	serine	tyrosine	cysteine	U
	phenylalanine	serine	tyrosine	cysteine	C
	leucine	serine	*punctuation*	*punctuation*	A
	leucine	serine	*punctuation*	tryptophan	G
C	leucine	proline	histidine	arginine	U
	leucine	proline	histidine	arginine	C
	leucine	proline	glutamine	arginine	A
	leucine	proline	glutamine	arginine	G
A	isoleucine	threonine	asparagine	serine	U
	isoleucine	threonine	asparagine	serine	C
	isoleucine	threonine	lysine	arginine	A
	methionine	threonine	lysine	arginine	G
G	valine	alanine	aspartic acid	glycine	U
	valine	alanine	aspartic acid	glycine	C
	valine	alanine	glutamic acid	glycine	A
	valine	alanine	glutamic acid	glycine	G

Figure 5.2. The genetic code.

of definiteness, here are a few of the code words from the ASCII code:

$$A = 1000001, \quad J = 1001010, \quad Q = 101001, \quad 8 = 0111000,$$

Fixed-length codes such as ASCII or the genetic code are reasonably efficient. But they are optimal only if the likelihood of every possible string (i.e., every possible message) is the same. In most natural languages, for instance, this is certainly not the case. The character "e" in English, for instance, occurs far more frequently than any of the other characters of the alphabet, about 10 percent of the time (which is why Ernest Vincent Wright's 267-page novel, *Gadsby*, published in 1939, is so, . . . , well, novel, because it contains not a single occurrence of this most-common letter!).

This leads one to conclude that a code using a shorter representation for frequently occurring characters may result in a shorter overall representation for large blocks of text. This is the idea underlying data compression. Let's illustrate the point.

One of the most successful plays on the London stage recently is a production called *Art*. The story revolves around a painting purchased for an outrageous price by the main character. What's odd about this painting is that it consists of a blank canvas painted entirely white. That's it. A totally white oil painting. As the friends of the owner come by to view this work, various opinions are expressed as to the artistic merit of such a work, ranging from "it's pure charlatanism and you were robbed" to "it's a masterpiece of understatement and a work of artistic genius." Regardless of the philosophical and artistic merit of this painting, the one thing that we could all agree on is that it is very easy to compactly describe this work: A canvas painted completely white. Here whatever information the entire canvas contains has been compressed into a phrase consisting of 35 characters in the English alphabet (including the space and period). And this compressed version suffices to faithfully recreate the original object. So not only is my phrase a compressed version of the original painting, it is uniquely decodable back into the original.

Unique decodability means that there can be only a single interpretation for each code. Roughly speaking, if a code is uniquely decipherable, it can't have very many short code words. For instance, if the word 010011 is a word of the code, then the words 01, 00 and 11, each of length 2, cannot be words, because if they were, then receipt of this longer string would leave open whether the message were the longer string or these three shorter substrings. That is, the string would not be *uniquely* decodable, because it could be decoded in more than one way. Let r be the number of symbols in the alphabet from which the code words are formed. For example, $r = 4$ for the genetic code because it is formed from the four symbols A, G, C, and T. In the mid-1950s

Brockway McMillan of Bell Labs proved the following result relating the notion of unique decipherability to the length of code words:

McMILLAN'S THEOREM   *Let the code words* $c_1, c_2, \ldots, c_q$ *of a code C have lengths* $\ell_1, \ell_2, \ldots, \ell_q$. *If C is uniquely decipherable these code-word lengths must satisfy the inequality*

$$\sum_{k=1}^{q} \frac{1}{r^{\ell_k}} \leq 1. \quad \blacksquare$$

What this result says (roughly) is that if you have a lot of symbols in your code (if r is large) then the code words can be longer (have bigger ℓs) and still be uniquely decipherable. The expression above makes this notion precise in a mathematical sense.

As an example, suppose we want to create a binary code whose code words are all of length not greater than 2. There are then six possible code words: 0, 1, 00, 01, 10, and 11. But this code is not uniquely decipherable because, for example, receipt of the code word 10 can be interpreted two different ways, either as the code word 1 or the code word 10. McMillan's Theorem can be used to show this same conclusion. In this case $\ell_1 = 1, \ell_2 = 1, \ell_j = 2, j = 3, 4, 5, 6$. Thus, the inequality to be tested is $1/2 + 1/2 + 1/4 + 1/4 + 1/4 + 1/4 = 2$, which is greater than 1. Thus, McMillan's result tells us that this code cannot be uniquely decipherable.

Note that McMillan's Theorem cannot tell us when a code is uniquely decipherable, only when it isn't. In other words, the theorem says that if a code is known to be uniquely decipherable, then its word lengths $\ell_1, \ell_2, \ldots$ must be such that they satisfy the inequality. The theorem is a necessary—but not sufficient—condition for unique decipherability. Also of importance for practical purposes is the quality that the code be *instantaneous*.

To illustrate this notion, consider the following two codes for the letters A, B, and C:

Message	Instantaneous Code	Noninstantaneous Code
A	0	0
B	10	01
C	11	11

Both of these codes are uniquely decodable. For example, upon receipt of the message 0111 using the noninstantaneous code, a decoder can work from right to left to determine that it is the message C followed by the message B.

Assume that the message 010110 is received. Working left to right with the noninstantaneous code, it's not possible to know whether message A was received or just the first half of message B. So with the noninstantaneous message it's not possible to decode the received message "on the fly," or instantaneously. But if the original message had been coded using the instantaneous code, then it can only be message A followed by B followed by C followed by A again; there is no ambiguity (unique decoding), and the message can be decoded as the code words appear without having to refer to earlier code words to pin down the message (instantaneously). The Morse code is an example of a code that is *not* instantaneous; the telegrapher must leave a letter space between each letter and a word space between words. So, in effect, the Morse code involves four symbols rather than simply just the symbols "dot" and "dash."

Given the importance of instantaneous codes, it is of considerable interest to have conditions that tell us when an instantaneous code exists. For example, consider the code $C = \{0, 11, 100, 110\}$. The number of code words is $q = 4$, and the code-word lengths are $\ell_1 = 1$, $\ell_2 = 2$, $\ell_3 = 3$ and $\ell_4 = 3$. Because the number of symbols in the code alphabet is $r = 2$, the inequality in Kraft's Theorem below is satisfied in this case because

$$\frac{1}{2} + \frac{1}{2^2} + \frac{1}{2^3} + \frac{1}{2^3} = 1 \leq 1.$$

But this code is definitely not instantaneous because the second code word is a prefix of the fourth one.

So we might ask: When does there exist an instantaneous code C with code words $c_1, c_2, \ldots, c_q$ having lengths $\ell_1, \ell_2, \ldots, \ell_q$? L. G. Kraft answered this question in 1949 with the following result:

KRAFT'S THEOREM *There exists an instantaneous code C with r symbols and code words $c_1, c_2, \ldots, c_q$ of lengths $\ell_1, \ell_2, \ldots, \ell_q$ if and only if these lengths satisfy the inequality*

$$\sum_{k=1}^{q} \frac{1}{r^{\ell_k}} \leq 1. \quad \blacksquare$$

Note carefully, though, that Kraft's Theorem says that if the lengths $\ell_1, \ell_2, \ldots, \ell_q$ satisfy the inequality then there must exist *some* instantaneous code with those code-word lengths. But it does not say that any code whose code-word lengths satisfy this inequality is necessarily instantaneous.

Kraft's Theorem may be used to create a specific instantaneous code with given codeword lengths. The full details can be found in the references for this chapter. Let's just give a flavor of how to proceed by the following example.

Assume we want to construct a binary code C that contains the code words 0, 10, and 110. Now, suppose we want to add some code words of length 5 to these code words. How many additional code words of length 5 can be added?

The three given code words contribute

$$\frac{1}{2} + \frac{1}{2^2} + \frac{1}{2^3} = \frac{7}{8}$$

to Kraft's inequality. So there is 1/8 left to work with. Any code word of length 5 will contribute $1/2^5 = 1/32$ to the sum. Thus, we can add at most four such code words. By examining the possibilities, we find that each code word of length 5 must begin with 111; otherwise, it would conflict with one of the previous code words. Thus, there are only four possibilities: 11100, 11101, 11110, 11111. It is not difficult to check that the code $C = \{0, 10, 110, 11100, 11101, 11110, 11111\}$ is indeed an instantaneous code.

We can put Kraft's Theorem and McMillan's Theorem together to conclude that if a uniquely decipherable code with code-word lengths

$\ell_1, \ell_2, \ldots, \ell_q$ exists, then an instantaneous code must also exist having these same code-word lengths. So we lose nothing by considering just instantaneous codes rather than the much larger set of uniquely decipherable ones. This is of great practical importance when it comes to the question of finding good codes with the shortest possible code-word lengths.

With these notions of codes and coding in hand, let's turn our attention to the first of the two big problems in information theory, the efficiency of a code.

Huffing and Puffing and Squeezing the Message Down

Probably the sine qua non of a good code is that we don't have two different code words standing for the very same message. This means that the code must be uniquely decipherable. McMillan's Theorem, however, shows that for this condition to hold, the code words must be fairly long. This requirement, on the other hand, works against the efficiency of the code by forcing longer strings to encode the messages.

But not all source messages, or symbols, occur with equal frequency. We saw earlier that in the English language, the symbol "e" occurs about 10 percent of the time. But the symbol "q" occurs with a frequency of only once in every 1250 characters, a probability of 0.0008. So it makes sense to think about a variable-length code in which the less frequent symbols are assigned longer code words than the frequently occurring ones. In this way, we might be able to improve the efficiency of our code.

We saw earlier that a source of messages is described by giving the symbols (messages) that might be transmitted together with the probabilities for these symbols. Suppose there are q symbols with probabilities $p_1, p_2, \ldots, p_q$. If C is a coding for these symbols with code words $c_1, c_2, \ldots, c_q$, then a reasonable measure of the efficiency of the code C is the *average code-word length,* given by the sum

$$c_1 p_1 + c_2 p_2 + \cdots + c_q p_q.$$

Coding I	Coding II
A → 11	A → 01010
B → 0	B → 00
C → 100	C → 10
D → 1010	D → 11

To illustrate this measure, suppose we have the source symbols A, B, C, and D with probabilities 2/17, 2/17, 8/17, and 5/17, respectively. Consider the two codings I and II above. By using the above probabilities, we can easily compute the average code-word length for coding I as:

$$\text{AvgCodILen} = 2(2/17) + 1(2/17) + 3(8/17) + 4(5/17) = 50/17,$$

whereas the average length for coding II is

$$\text{AvgCodIILen} = 5(2/17) + 2(2/17) + 2(8/17) + 2(5/17) = 40/17.$$

So we find that coding II has a shorter average code-word length than coding I, which means it is more efficient than coding I.

This example leads to the obvious question: What is the *most* efficient coding scheme, the one with the smallest average code-word length, for a given source? This question was answered by D. A. Huffman in 1952, who discovered an ingenious scheme for creating optimal codes. This method, now called *Huffman encoding,* is of the utmost importance in coding theory for the reasons stated in the following theorem.

HUFFMAN'S THEOREM *Let S be an information source. Then Huffman encoding of S is instantaneous. Moreover, Huffman encoding has the smallest average code-word length among all instantaneous encoding schemes for S.* ∎

The essential idea in Huffman's procedure is to systematically assign the shortest code words to the symbols that occur most frequently, and to do so in such a way that the code obtained is instantaneous. An example of Huffman encoding due to John Pierce of Bell Labs, one of

Word	Probability
the	0.50
man	0.15
to	0.12
runs	0.10
house	0.04
likes	0.04
horse	0.03
sells	0.02

the pioneers of information theory, for a set of words ("symbols") in English graphically illustrates Huffman's method. Suppose the words to be coded and their respective probabilities to be as above.

Construction of the Huffman code for this source proceeds according to the following steps:

1. Find the two smallest probabilities, 0.02 (*sells*) and 0.03 (*horse*), and draw a line from each of them to their sum, the point 0.05, which is the probability of either the symbol *sells* or the symbol *horse* being transmitted. This step is shown at the bottom of Figure 5.3.

2. Now look for the two lowest probabilities among all the individual *and* compound events (points corresponding to the fusion of two symbols, such as *sells* and *horse,* as above) that remain. These happen to be 0.04 (*like*) and 0.04 (*house*). Draw lines from them to the right to a point marked 0.08, the sum of these two probabilities.

3. The two lowest remaining probabilities are now 0.05 (the sum found earlier) and 0.08 (the sum just obtained). So draw lines connecting them, to give a point marked 0.13. Continue in this fashion by finding at each step the two lowest probabilities remaining, until all paths run from each word to a common point at the right marked 1.00.

4. Put a 1 on each upper path going from the left and a 0 on each lower path.

227

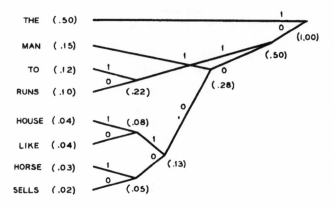

Figure 5.3. A Huffman code for an English-language source.

5. The Huffman code for a given word is then the sequence of digits encountered in tracing back from the common point 1.00 to the word in question.

The diagram obtained by following the foregoing steps 1–5 is shown in Figure 5.3. In this example, to reach the word "like," we go down one step from 1.00 along a line labeled 0. From there, we go down one more step along another line labeled 0. Carrying on this process to reach "like," we pass along three down steps, followed by an up step and another down step, leading to the code word 00010.

The Huffman code for the various words in this source is

house = 00011,   like = 00010,   horse = 00001,
sells = 00000,   the = 1,   man = 001,   to = 011,   runs = 010.

The Huffman code is a good example of a nearly optimal code guaranteed by Shannon's Theorem. Let's now turn to a consideration of Shannon's Theorem itself.

In the opening section of this chapter, we saw that a measure of the amount of information content in a source S is its entropy, expressed as

$$H(S) = - \left[P(s_1) \log P(s_1) + P(s_2) \log P(s_2) + \cdots + P(s_n) \log P(s_n) \right],$$

where s_k is the kth symbol in the symbol source and $P(s_k)$ is its probability. By calculating the entropy of the source of eight symbols using the probabilities given previously in the table, we find $H(S) = 2.21$ bits per word. Calculating the average length of a word in the Huffman coding yields 2.26 bits per word. This should be compared with the 3 bits per word that would have been needed to code these eight symbols if we had simply used the eight possible 3-digit binary code words.

When encoding a source, it seems reasonable to suppose that we will need at least as many bits of information in the encoding as is in the source. After all, one can hardly expect to be able to reconstruct the message if the coding scheme destroys information in the source! But commonsense supposition and provable fact don't always coincide. Fortunately, though, in this case, supposition was turned into fact by Shannon, who showed that the smallest average code length must be at least as great as the entropy in the source. More formally, we have Shannon's famous Noiseless Coding Theorem.

SHANNON'S NOISELESS CODING THEOREM   *Let S be an informa-tion source with entropy* H(S). *Then*

$$H(S) \leq \text{MinAveCodLen}\,(S).$$

Here MinAveCodLen (S) *denotes the minimum average code-word length among all uniquely decipherable coding schemes for* S. ∎

Shannon's Noiseless Coding Theorem in the above form provides a lower bound on the best possible code-word lengths; the code-word lengths must be at least as long as the entropy of the source, on the average. Some algebraic playing about with Kraft's inequality shows that an upper bound is given by

$$\text{MinAveCodLen}(S) < H(S) + 1.$$

From this we see that the Huffman code always calls for less that one bit per symbol more than the entropy. In the example given a moment ago, the Huffman coding requires just 0.05 bits per word more than the entropy of the source. Even this tight fit can be improved with more esoteric—but more difficult to construct—codes. But it would take us a bit far afield to go into the details here. The motivated reader can find them in the chapter references.

So far, we have concentrated on situations in which there is no noise—in the encoding process, in the communication channel, or in the decoding. So efficiency in the coding scheme has been our goal. This is all right for getting a feel for the limits of what's possible by way of encoding and transmitting information. But when you start talking about real-world communication systems, you're talking about noise in the source, the communication channel, and maybe even in the receiver. So it's time to consider in detail Problem 2, sketched earlier, in which we ask about the maximum amount of information that can be transmitted over a noisy channel with a given rate of error.

Symbols, Signals, and Noise

In 1948, the MIT mathematician Norbert Wiener published his famous book *Cybernetics,* which treated (among many other things) the problem of extracting signals of a given class from noise of a known type. In the very same year, Claude Shannon of Bell Labs published a paper in two parts that is regarded as the foundation of information theory. Shannon's work addresses the question of how to encode messages chosen from a known ensemble, so that they can be transmitted accurately and quickly in the presence of noise. So while Wiener was focused on the problem of the receiver extracting an untreated signal from noise, Shannon emphasized the problem of how the sender could process the message before transmitting it, in order to make the message clear to the receiver.

The common element faced by both Wiener and Shannon is the noise in the communication channel. In telephone lines and radio circuits we hear the desired signal against a background of crackling static and "hissing." Television pictures are overlaid with granular "snow," and in binary data transmission the received character may sometimes be "flipped" from the character that was sent. An engineer faced with these problems 50 years ago would probably have advised either increasing the power in the transmitter or introducing some sort of redundancy in the message by resending it sufficiently often. Shannon offered another solution.

The basis of Shannon's solution to the noisy channel problem is to tell the engineer that by properly encoding the message, it can be sent over even a noisy channel with an error rate below any predefined level.

Essentially, Shannon's famous Coding Theorem can be paraphrased as follows: "If both the noise in the channel and its capacity for transmission are known, then I can calculate how many characters can be sent over the channel per second, and if I send any number of characters less than this I can do so virtually without error. But if I try to send more, errors will necessarily creep in." This is the Shannon Coding Theorem, one of the most surprising results in modern engineering. Understanding why this result is *inevitable,* rather than surprising, is one of the principal goals of this chapter.

Earlier, we saw that the capacity of a communication channel to transmit information is expressed as the difference between the information contained in the source, S (the entropy $H(S)$) and the information contained in the received signal at the destination, D, given the particular source (the conditional entropy, $H(S|D)$). We have already seen that noise is the only barrier to perfect communication. So let's see how the presence of noise affects this conclusion.

Suppose at every instant of time $t = 1, 2, \ldots, T$, the source sends a symbol. Moreover, let there be a noise element n_t present in the channel at each of these instants. If we denote the sequence of noise terms by $N = \{n_1, n_2, \ldots, n_T\}$, then the conditional entropy $H(S|D)$ becomes the entropy of the noise, call it $H(N)$. This simply means that the only thing standing in the way of the receiver knowing exactly what is transmitted is the channel noise N. Thus, the channel capacity becomes

$$C = \max[H(S) - H(N)], \qquad (*)$$

where the maximization is taken over all possible coding schemes for the source. In other words, take every possible coding scheme and computer $H(S)$ for that scheme. Then use the scheme that gives the largest value for $H(S)$. Let's now look at how bad the noise term $H(N)$ can be. We want to find the *type* of noise that will make $H(N)$ as large as possible, so that we can get a worst-case estimate on the capacity of the channel.

In general, we can write $H(N)$ as

$$H(N) = H(n_1, n_2, \ldots, n_T),$$

which is less than or equal to

$$H(n_1) + H(n_2) + \cdots + H(n_k),$$

with equality only when the noise variables n_i are statistically independent. This is because in the case of independence, the noise n_i at time i

has no influence on the noise at any other time j. The worst case then occurs when

$$H(N) = \sum_{k=1}^{T} H(n_k).$$

So we have to look for a probability distribution that maximizes $H(n_i)$. Cutting to the chase, this problem was solved years ago in the engineering literature. The maximum is attained when each noise term n_i is distributed in accordance with the famous bell-shaped curve of the normal distribution. We conclude, then, that the worst case arises when all the noise components n_i are independent and obey the normal distribution.

What Shannon's Noisy Coding Theorem shows is that the signaling rate can approach as close as one likes to the channel capacity, keeping the accuracy arbitrarily high.

We return now to the problem of finding the coding scheme of the source that maximizes expression (∗). It turns out that the optimal coding of the source is such that if we let Γ represent the so-called *signal-to-noise* ratio in the source, the ratio of the average energy in a sample of the signal to the average energy of the noise, the channel capacity then becomes

$$C = \log(1 + \Gamma) \text{ bits per second.} \tag{†}$$

So for large Γ (strong signal and/or weak noise), C is approximately equal to $\log \Gamma$, whereas for a small Γ (weak signal, large noise), the capacity is approximately equal to Γ.

Expression (†), which is known as Shannon's formula, says that the channel capacity remains larger than zero, even when the signal is much weaker than the noise. Because the expression for the channel capacity is for the worst possible case of noise, the fact that the capacity remains above zero is what underlies Shannon's Coding Theorem.

The foregoing result provides only an estimate of how large the channel capacity can be. Generally speaking, it is difficult to compute the channel capacity precisely, one reason being that the capacity depends on the type of noise in the channel, which may vary greatly. But one case in which the capacity can be calculated exactly is for the so-called *symmetric binary channel*. This is a channel in which only the symbols 0 or 1 are transmitted, and in which the likelihoods of an "inversion" in which 0 is sent and 1 is received, or vice versa, are the same. For such a channel, maximum information transfer will occur when the probability of sending a 1 equals the probability of sending a 0. Under these circumstances,

$$H(S) = H(D)$$
$$= -\left(\frac{1}{2}\log\frac{1}{2} + \frac{1}{2}\log\frac{1}{2}\right)$$
$$= 1 \text{ bit per symbol.}$$

For the conditional entropy $H(S|D)$, we have

$$H(S|D) = -p\log p - (1-p)\log(1-p),$$

where p is the *crossover probability* of incorrect transmission. That is, p is the probability that if 1 was sent, 0 was received (and similarly for sending 0 and receiving 1). Thus, we find the channel capacity of this symmetric binary channel to be

$$C = H(S) - H(S|D)$$
$$= 1 + p\log p + (1-p)\log(1-p)$$

From this analysis, note that if $p = \frac{1}{2}$, the channel capacity is 0. This makes sense, because if we receive a 1 it is equally likely in this case that a 1 or a 0 was transmitted. So the received message does nothing to reduce our uncertainty as to what symbol was sent. Note also that the capacity is the same when $p = 0$ or $p = 1$ because if we consistently receive a 0 when we send a 1 (or a 1 when we send a 0) we are as certain of what was transmitted as if the channel had not corrupted the signal at all!

Now we can state Shannon's pioneering result:

NOISY CODING THEOREM (FOR SYMMETRIC BINARY CHANNELS)
Suppose we have a channel with crossover probability p *and capacity* C *(depending on* p*). If the transmission rate* R *of a code is less than* C, *then for any* $\epsilon > 0$, *there exists a code* C *whose transmission rate is no less than* R, *and for which the probability of a decoding error is less than* ϵ. ∎

Roughly speaking, this result says that if we choose any transmission rate below the channel capacity, there exists some code that can transmit at that rate while maintaining an arbitrarily small probability of error at the destination.

There are two aspects worth commenting upon. First, the price to be paid for this essentially error-free transmission is that the code-word lengths may have to be very large. Second, Shannon's result is an existence theorem; it says that such a "good" code must exist. But it says nothing whatsoever about how to find it. So the good news is that there is a needle in the haystack; the bad news is that we have to use other means to dig it out. Let's turn now to some of the means that people have used to find explicit codes that perform the trick that Shannon's theorem says can be done.

If It Ain't Broken, Fix It Anyway

Suppose an unscrupulous gambler in Las Vegas wanted to bet on a horse race taking place in, say, London. To make things simple, suppose the race involves only two horses, Sir Winston and Lord Nelson. The gambler arranges with a confederate in London to send a coded message of the outcome of the race because the betting shops in Nevada are 8 hours behind events in Great Britain. The symbol 1 means that Sir Winston was the winner of the race, and 0 means that Lord Nelson romped home triumphant. The gambler's problem is that the transatlantic phone lines are less than noise free, and he wants to make sure of the result before placing a big wad of cash in Vegas on one of these nags or the other. The simple length one code words 0 and 1 are far too likely to be corrupted to trust, so the gambler has to create a better code that will compensate for possible errors. How to do it?

One way to generate error-correcting codes is to send code words that differ from each other so much that an error cannot change one of the code words into another. In the gambler's case, where the simplest code has just the two code words, 0 and 1, no errors can be detected because any error changes one codeword into the other. But the new code {00, 11}, formed by simply repeating the digits of the old code words, will detect single errors because any single error in a code word transforms it into a noncode word. For example, 00 could be sent and

01 received. The gambler would then know that there was something wrong with the transmission, and could ask for a resend. But notice that a transmission that contained *two* errors rather than one could not be detected by this scheme.

By repeating each of the binary symbols, 0 and 1, $n - 1$ times, we arrive at what's called the *repetition code,* consisting of two code words of length n. These two words are $000 \cdots 000$ and $111 \cdots 111$. Up to $n - 1$ errors in transmission are detectable by this coding, where the first digit is the message and the remaining $n - 1$ digits are the check digits. This elementary scheme allows the sending of only two different code words, but can be easily extended by using repetition of blocks rather than just single digits. Details can be found in the chapter references.

As we have repeatedly emphasized, the goal of coding theory is to concoct ways to send messages quickly and accurately. Achieving accuracy over a noisy communication channel requires more error detection, which means more check digits per code word. Thus, longer code words are required, which in turn entails slower transmission of information. The repetition codes have high reliability, and hence high accuracy, but they are very slow. The notion of using the check bits for what's called *parity checks* for errors turns out to help realize goals, speed, and reliability. Let's look at a couple of examples.

- *Universal Product Code:* With cheap, reliable, and fast laser scanning devices and computers, it is now standard practice in most food stores to attach "zebra stripe" product identification labels to items for pricing, inventory control, and faster checkouts. Probably the most familiar coding scheme of this sort is the Universal Product Code (UPC) for grocery items. This is a code involving ten symbols, the digits 0, 1, 2, . . . , 9, arranged in blocks of length 11. So there are 10^{11} possible code words—sufficient for even the most well-stocked market, which might carry at most 30,000 or so separate items. A typical UPC codeword is shown in Figure 5.4.

A number $a_{11}a_{10} \ldots a_2a_1$ identifies the particular product. To detect errors, a check digit a_0 is added to the right so that the number

$$3a_{11} + a_{10} + 3a_9 + a_8 + \cdots + a_2 + 3a_1 + a_0$$

is a multiple of 10 because the number base used in the bar code is the usual decimal system (not the binary system we have used so far in this chapter). In the code word shown in Figure 5.4, which is the

Figure 5.4. Example of a UPC code word.

number 07507031400, it is easy to calculate the above sum; it is 65. Therefore, the check digit 5 should be added to bring this sum to a number divisible evenly by 10. You can see from the figure that this has indeed been done. The check digit in this case is called a parity digit. A bit of thought shows that an error in a single position would result in a sum that is not a multiple of 10, and thus would be detected. Similarly, this parity check would detect the transposition of adjacent digits x and y—provided that the difference between the two digits is not equal to 5, the parity digit itself.

• *International Standard Book Number:* All publishers in the Western world code each of their books upon publication with a scheme involving the same 10 symbols as the UPC, but with code words of length 9. This is the so-called International Standard Book Number, or ISBN. Each publication has a 9-digit identification number $a_9 a_8 \ldots a_1$ to which the parity digit a_0 is appended so that the number

$$10a_9 + 9a_8 + 8a_7 + \cdots + 2a_1 + a_0$$

is a multiple of 11. (*Note:* If a_0 needs to be 10 to get this sum to be a multiple of 11, X, the roman digit for 10, is used.) This coding scheme detects all single-position errors, as well as transpositions, but employs a nonnumeric character.

The UBC and ISBN codes are examples of *error-detecting* codes. But in and of themselves they do nothing about correcting any errors discovered. They simply serve as red flags to indicate that something has gone amiss. It's one thing, however, to know that a message is wrong; it's quite something else to be able to reconstruct the correct message from the wrong message. This leads to the notion of *error-correcting* codes.

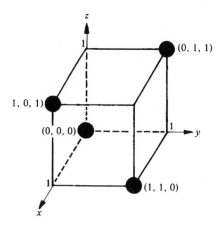

Figure 5.5. Example of a binary code with even parity.

To look at the basic idea underlying construction of error-correcting codes, consider a cube in three-dimensional space. In the usual $x - y - z$ coordinate system, the vertices of this cube can be associated with the eight binary digits of length 3. A binary code consisting of code words of length 3 is then simply a subset of the vertices of this cube. For example, Figure 5.5 shows geometrically a code with four code words of length 3—the words 000, 011, 101, and 110—which have been formed by adding a parity check digit to the end of each of the binary digits of length two: 00, 01, 10, and 11. The parity digit in this situation just ensures that the sum of the digits in each code word sums to 2 (even parity). The figure shows the code words as the darkened vertices of the unit cube.

In Figure 5.6, we see the three possible results of a single error made in transmitting the code word 101. The single error changes one coordinate, which means that it can be detected as long as no other codewords are a single coordinate away from 101. The figure shows that this is indeed the case because the blackened vertices are all at least two coordinates away from 101. Thus, if we were to receive any of the three blocks, 001, 111, or 100, then there must have been an error in transmission. Examination of the figure shows the same argument can be made for a single error in any of the other four code words. So this code can detect an error if there is but at most a single error in transmitting the digits of the block representing the message. What about error correction?

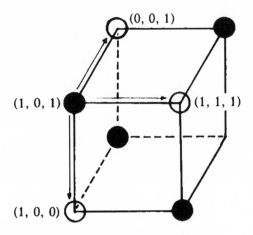

Figure 5.6. The effect of a single error in the code word 101.

The preceding argument served to motivate the work of Richard W. Hamming of Bell Labs in the late 1940s, who defined what is now called the *Hamming distance, D,* between two blocks of n binary digits. In terms of the geometry of the unit cube, this distance is just the number of edges in a shortest path between the two vertices. The distance $D(x, y)$ between two binary digits, x and y, is just the number of coordinates at which the two digits differ. The minimum distance of a code is then the minimum of all distances between any two distinct code words of the code. For the code shown in Figure 5.5, this minimal distance is two.

In terms of the Hamming distance, single-error detection is possible with a code whose minimum distance is two. But single-error correction is possible with a code whose minimum distance is at least three. An example of this is given by the code depicted in Figure 5.7. Here the code words are just the two blocks 111 and 000.

Figure 5.8 shows the words that might be received if a single error is made in transmitting the code word 000. Each of the three possible received words is a distance two away from the code word 111. So if at most one error occurs in the transmission of 000, then the error can be found and we can figure out exactly which code word has been sent. For instance, if we receive the word 001, it can only be a corruption of the code word 000 because we have assumed that at most one error can occur. So the received word cannot be an error in 111, as that code word is a distance two away from the received word—one unit too far to be

238

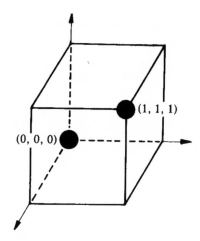

Figure 5.7. An error-correcting code.

a candidate for the correct code word. But, of course, if we allow the possibility of *two* errors in transmission, all bets are off, and this code cannot correct such errors (but it can detect them).

This example suggests a general rule for error correction. It is simplicity itself: Whatever word you receive, go to the nearest code word in the unit cube measured in the Hamming distance.

As a further illustration, take the code consisting of the two code words 00000 and 11111. Suppose 10100 is received. The receiver argues that it's more likely that 00000 was sent with two errors than that 11111

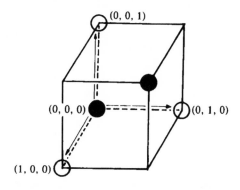

Figure 5.8. Possible received words from a single error in 000.

was sent with three mistakes. Therefore, the receiver decodes 10100 as 00000. So what the receiver has done is to decode the received word as the closest code word measured by the Hamming metric.

In general, if d is the minimum Hamming distance between code words, the code can detect up to $[d/2]$ errors and simultaneously correct $e = [(d-1)/2]$ errors, where the square brackets denote the greatest integer function. So if x is some real number, $[x]$ is the largest integer that is still smaller than x. Some of the correction capability may be sacrificed for additional detection capability. For instance, a code with minimum distance d can detect up to $d-1$ errors by foregoing all correction.

Certainly the most well-known error-correcting code is one devised by Hamming and described by Shannon in just a few lines in one of his famous 1948 papers. This is the *(7, 4) code,* and involves each received word being subjected to a system of three parity checks that taken together produce a three-digit checking number. The following is the description of the (7, 4) code in Shannon's own words:

> Let a block of seven [binary] symbols be $X_1, X_2, \ldots, X_7$. Of these X_3, X_5, X_6 and X_7 are the message symbols and chosen arbitrarily by the source. The other three are redundant and calculated as follows:
>
> X_4 is chosen to make $\alpha = X_4 + X_5 + X_6 + X_7$ even
> X_2 is chosen to make $\beta = X_2 + X_3 + X_6 + X_7$ even
> X_1 is chosen to make $\gamma = X_1 + X_3 + X_5 + X_7$ even
>
> When a block of seven is received, α, β and γ are calculated and if even, called zero, if odd, called one. The binary number $\alpha\beta\gamma$ then gives the subscript of the X_i that is incorrect (if 0, there was no error).

Shannon presented this code as an example of one whose transmission rate exactly matched the capacity of a certain channel. The channel was to consist of blocks of binary digits of length 7, each affected by noise in at most one location. A diagram of this channel is shown in Figure 5.9. The capacity of such a channel is 4 bits per block, which amounts to 4/7 = 0.57 bits per digit—exactly the transmission rate of Hamming's (7, 4) code.

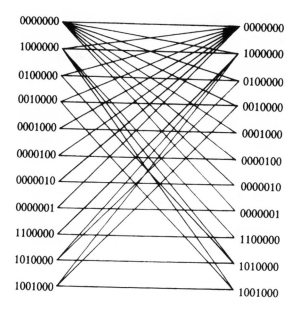

Figure 5.9. The channel for Hamming's (7, 4) code.

For ease of writing, let's call a binary code of minimum distance d whose code words are of length n and that contains M code words, an (n, M, d) code. Then perhaps the best measure of the error-correcting capability of a code C is not the absolute number of errors that it corrects, $[(d - 1)/2]$, but rather the *relative* number of errors that it can correct per code-word position. Correcting, say, 1,000 errors per code word sounds pretty good—until you discover that the code-word length is a billion! So we define the *error correction rate* of an (n, M, d) code to be $\delta = [(d - 1)/2]/n$. To illustrate this definition, consider the Hamming $(n, 4)$ code, which has $M = 2^{n-3}$ code words. The transmission rate of this code is $1 - 3/n$, which is excellent for n large. But the error-correction rate is just $\delta = 1/n$, which tends to zero as n becomes large. This same conflict between fast transmission and accurate transmission arises with all other codes. What all this adds up to is that you can't have it both ways; a compromise between transmission rate and quality of error correction has to be found. Coding theorists are still seeking that "sweet spot" that balances out these two competing goals. The reader should look at the chapter references for a full account of the search for this Holy Grail of coding. For now, let's return to the genetic code shown earlier in Figure 5.2, and look at how some of the basic concepts of information theory can be applied to its investigation.

The Information of the Genes

Shortly after Francis Crick and James Watson of Cambridge University discovered the double-helix structure of the DNA molecule in 1953, Crick, Sydney Brenner, also at Cambridge, and others put together the genetic code shown in Figure 5.2. This is the code by which the nucleic acid pattern written onto the DNA strand is mapped into the amino acid sequences that make up an organism's proteins, the real "do'ers" of life. To make it as explicit as possible that this "code" really is a code, Figure 5.10 shows the correspondence between the generic communication system of Figure 5.1 and the system used by the cellular machinery to get from the DNA to proteins.

The first step in mapping between DNA and the characteristics of the organism is that the order of the bases in the DNA determine the order of the amino acids in the protein chain that is synthesized by the cell. So, how many nucleotide bases are necessary to determine that a particular amino acid should be added to a chain being synthesized? There are four

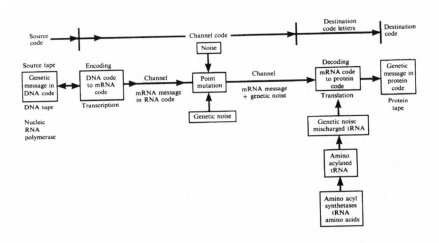

Figure 5.10. The genetic communication system.

242

nucleotide bases, so they could determine the formation of four different amino acids. But nature uses 20 amino acids, so this is not enough. If pairs of nucleotide bases are used, then there are $4 \times 4 = 16$ amino acids that could be coded for with such a "doublet" code—still not enough. Thus, only combinations of three nucleotide bases are sufficient to form the 20 different signals that correspond to the 20 amino acids.

In information-theoretic terms, we can say that a nucleotide carries $\log 4 = 2$ bits of information. But an amino acid carries $\log 20 = 4.32$ bits. So if the nucleotides are to determine the amino acids, each nucleotide "word" must contain at least this much information. Combinations of two nucleotides contain only $\log 4^2 = 4$ bits, which is not quite enough. But since $\log 4^3 = 6$ bits, which is greater than 4.32 bits, we find that a triplet code is the smallest that will do the job. But such a code of 64 words has a lot of extra words left over, and thus seems far from optimal in a purely information-theoretic sense.

Biochemical research has shown that the "extra" nucleotide triplets play several different roles. First, examination of Figure 5.2 shows that there is some redundancy in the genetic code. This serves to partially protect the integrity of the code against genetic noise. Some of the other nucleotide triplets, however, play a very different role, in that they don't code for amino acids at all. Rather, they serve as punctuation marks, telling the protein-manufacturing machinery when the job is done. So in this sense they act as the period at the end of this sentence, which in written language serves to terminate one thought and begin another.

We can also compute the channel capacity of the genetic system from the information transfer

$$I(S \rightarrow D) = H(S) - H(S|D).$$

Here S is the source code words, the triplets of nucleotide bases, known as *codons*. The destination D is the set of possible amino acids. The maximum value of $H(S)$ occurs when all codons are equally probable, in which case the probability of each is 1/64. This means that the nucleotide bases should be present in equal amounts, which is known to be approximately the case for the bacteria *E. coli*. However, the situation is different in some other cases, so there is no universality in the value of I. But leaving this aside, to a first approximation the genetic system source S has 64 symbols and the number of states of the destination D

is 23 (the 20 amino acids, plus three symbols for punctuation, start, and stop).

With these figures, the channel capacity can be taken as $H_{\max}(D) = \log 23 = 4.52$ bits per amino acid. The redundancy of the channel is then at least $1 - H_{\max}(D)/H_{\max}(S) = 1 - (4.52)/6 = 25$ percent. It should be noted that this figure decreases as the cell ages because information is lost due to random mutations.

Turning to the information-theoretic properties of the genetic code itself, we find that the code has two extremely important properties. It is both instantaneous and optimal.

• *Instantaneous:* Recall that a code is instantaneous if no code word is the prefix of any other code word. Suppose the genetic code evolved from a simpler doublet code, when the number of amino acids used by nature increased beyond 16. For the doublet code to be instantaneous, Kraft's inequality is a necessary and sufficient condition for that primitive genetic code to be instantaneous. In the case of q code words using an alphabet of r symbols forming words of length ℓ, this inequality is

$$\sum_{i=1}^{q} r^{\ell_i} \leq 1.$$

For the doublet genetic code, $r = 4$, $q = 16$, and $\ell = 2$. Plugging these values into the above expression, we find the Kraft sum to be

$$\sum_{1}^{16} 4^{-2} = 16 \times \left(\frac{1}{16}\right) = 1.$$

Thus, Kraft's inequality is satisfied and the doublet code is instantaneous. For the triplet code of today, the above numbers are $r = 4$, $q = 64$, and $\ell = 3$. It's an easy computation to see that Kraft's inequality is also satisfied by today's genetic code; hence, it is instantaneous.

The fact that the genetic code is instantaneous is of great practical importance because if it were not, some kind of storage mechanism would be needed to record the message before decoding could proceed. This is because we would not be able to decode the message character-by-character as the characters arrived, but would have to wait for the entire string to appear before decoding could commence.

• *Optimal:* An optimal code is one that is both instantaneous and has the minimum average code-word length. It turns out that the genetic code does indeed make the most economical use of its nucleotide symbols. If we have an information source S whose entropy is $H(S)$, then the maximum compression of a code for this source is just $H(S)/\log q$, where q is the number of symbols in the code's alphabet ($q = 4$ for the genetic code). For the worst case when all codons are equally likely, it can be shown that the above ratio gradually increases from 2 to 3 as the code moves from a doublet code to its modern word length of 3. So the genetic code is optimal.

There are many other fascinating aspects of the genetic code, but we have no space to enter into them here. Such questions ask whether the genetic code evolved in such a way that similar amino acids have similar codons, or whether it's really the case that only 20 amino acids are directly incorporated in the process of protein synthesis. These and many additional conundrums are discussed in detail in the chapter references.

Randomness, Information, and Computation

Shannon's notion of the information content of a string of symbols involves looking at an ensemble of messages in which the symbols making up the message appear with different probabilities. Thus, right from its inception, information theory contained an inherent notion of randomness, or uncertainty, in that the symbol appeared only in a statistical sense. About a decade or so after Shannon's pioneering work, the American Ray Solomonoff and the Russian Andrei Kolmogorov, and most significantly, Gregory Chaitin of IBM, began to develop the idea of what is now termed *algorithmic information theory (AIT)*. The motivation for this work was very different from the problems of communication that stimulated Shannon's investigations. These AIT researchers had in mind the problem of measuring the information content of a particular string of symbols by how "complex" the string might be. They approached the question via the notion of a computer program that would produce the given pattern of symbols. But while the motivations were very different between information theory (IT) and AIT, it turns out that the two theories have many important points in common. And it helps illuminate many aspects of information theory to look at the concepts from a computational point of view.

In the early 1960s, Chaitin enrolled in a computer-programming course being given at Columbia University for bright high-school students. At each lecture the professor would assign the class an exercise that required writing a program to solve it. The students then competed among themselves to see who could write the shortest program solving the assigned problem. This spirit of competition undoubtedly added some spice to what were otherwise probably pretty dull programming exercises, but Chaitin reports that no one in the class could even begin to think of how to actually prove that the weekly winner's program was really the shortest possible.

Even after the course ended, Chaitin continued pondering this shortest-program puzzle, eventually seeing how to relate it to a different question: How can we measure the complexity of a number? Is there any way that we can objectively claim π is more complex than, say, $\sqrt{2}$ or 759?

In 1965, while an undergraduate at the City University of New York, Chaitin had the bright idea to define the complexity of a number to be the length of the shortest computer program that will cause the machine to print out that number. Using this idea, he came up with the following complexity-based definition of a random number: A number is *random* if the shortest program for calculating the number is not appreciably shorter in length than the number itself. Expressing this another way, we can say a number is random if it is maximally complex. Here, of course, we take the length of a number or program to be the number of binary digits needed to write down that number or program. With this definition, a number such as $\pi = 3.14159265\ldots$ is not random because arbitrarily many digits of π can be generated using any of a number of known programs of fixed length. Nevertheless, an infinitely long number such as π is certainly more complex than a simple number of finite length such as 47 because we can always use the program "PRINT 47" to generate the latter quantity. This shortest program for 47 is quite a bit shorter than the shortest program that will successively crank out the digits of π.

So if something as complicated looking as π isn't random, do random numbers really exist? Or does Chaitin's definition define an empty set? The surprising fact is that by this information-theoretic view of what is random, almost *all* numbers are random! To see why, think about what it means for a number to be nonrandom. By definition, a number is nonrandom if it can be produced by a computer program whose length is significantly shorter than the length of the number itself. Suppose we consider all numbers having length n, that is, all binary strings of n digits. Each of the n digits can be either 0 or 1, so there are a total of 2^n numbers of length n. Let's compute the fraction of these numbers having complexity less than, say, $n - 5$. That is, let's look for all numbers of length n that can be produced by a computer program that can be coded in no more than $n - 5$ bits.

Because our interest is in all computer programs that can be coded in no more than $n - 5$ bits, we could actually list each one of these programs. So, for example, there is one program of length zero (the empty program consisting of no instructions), two programs of length 1 (the single-element strings 0 and 1), four programs of length 2 (the strings 00, 01, 10, and 11) and, in general, 2^k programs of length k. Counting all possibilities, there is a total of $1 + 2 + 4 + \ldots + 2^{n-5} = 2^{n-4} - 1$ such programs having length $n - 5$ or less. Consequently, there are at most this many numbers of length n whose complexity is less than or equal to $n - 5$ because at best each of these programs can produce an output corresponding to an actual number of length n. But we have seen that there is a total of 2^n numbers of length n. Therefore, the proportion of these numbers having complexity no greater than $n - 5$ is at most $(2^{n-4} - 1)/2^n \leq 1/16$.

Therefore, no more than one number in 16 can be described by a program that's at least 5 bits shorter than the number itself. Using this same argument, we can show that no more than one number in 500 can be produced by a program 10 or more bits shorter than the number's length—that is, its complexity is 10 or more units away from being random. Continuing in this vein and letting $n \to \infty$, we can fairly easily prove that the set of real numbers having less than maximal complexity forms an infinitesimally small subset of the set of all numbers. In short, almost every real number is random because there exists no program that produces the number that is shorter than the trivial program that just prints the number itself. Now let's look at this shortest program business in a little more depth.

The implication of Chaitin's Theorem is that, for sufficiently large numbers N, it cannot be proven that a particular string has complexity greater than N. Or, what is the same thing, there exists a level N such that no number whose binary string is of length greater than N can be proven to be random. Nevertheless, we know that almost every number is random. We just can't prove that any *given* number is random.

Chaitin's Theorem says that if we have some program, there always exists a finite number t such that t is the most complex number our program can generate. Nevertheless, we can clearly see that numbers having complexity greater than t do exist. To construct the binary string for one, simply toss a coin a bit more than t times, writing down a 1 when a head turns up and a 0 for tails. When we start talking about randomness and strings of zeros and ones, information theory can't be far away. So let's make the connection between IT and AIT.

Consider a prototypical situation in which there is a source S that can transmit three possible symbols, call them A, B, or C. Suppose the source sends these symbols with probabilities 1/6, 1/6, and 2/3, respectively. Then the entropy of S is defined by

$$H(S) = - \left(\frac{1}{6} \log \frac{1}{6} + \frac{1}{6} \log \frac{1}{6} + \frac{2}{3} \log \frac{2}{3} \right).$$

So to encode a typical message from this source that is N symbols long, $N \times H(S)$ bits will be required.

In Chaitin's version of information theory, one looks at individual long messages, not statistical *ensembles* and average number of bits, as in Shannon's version. According to the Law of Large Numbers, a long message of N characters from the source S will generally have about $N/6$ As, $N/6$ Bs, and $2N/3$ Cs. Chaitin then observed that if N is large, such a message may be compressed into a computer program that is about $N \times H(S)$ bits in length that will generate the message.

Two basic quantities in information theory that we have encountered several times in this chapter are the formula for the entropy of a single symbol X,

$$H(X) = - \log P(X), \tag{§}$$

where $P(X)$ is the probability of the symbol X occurring, and the formula for the mutual entropy of symbols X and Y,

$$H(X, Y) = H(X) + H(X|Y). \tag{¶}$$

In Shannon's theory, these formulas are *definitions,* made to exploit certain very convenient properties of the logarithm function.

These *identical* expressions appear in AIT as well—but with the notable difference that in AIT they are *theorems,* not definitions. In Chaitin's theory of program-size complexity, formula (§) states that the size of the smallest program to compute some quantity X is (up to an additive constant) the logarithm of the probability that a program chosen at random calculates X. Formula (¶) means that the smallest program that calculates two quantities X and Y is (again, up to an additive constant) equal to the sum of the size of the smallest program that calculates X and the size of the smallest program that calculates Y, given the smallest program for computing X.

As an interesting aside, the first of these fundamental formulas, (§), implies that most of the probability for computing a quantity X is concentrated in the minimum or near the minimum-size programs for computing X. It also implies that there are very few minimum-size programs for calculating any quantity.

Zipf Up That Lip

In 1949, just one year after Shannon published his pathbreaking work establishing the field of information theory, linguist George Zipf published the book, *Human Behavior and the Principle of Least Effort.* In this volume, Zipf announced an empirical relationship between the rank order of words in a language such as English, and the frequency of their occurrence. This relationship, now termed *Zipf's law,* is very closely related to the principles laid down by Shannon. So it's fitting to conclude this chapter with an account of how Zipf's law can be derived directly from information-theoretic reasoning.

Suppose we list the words of the English language according to how common they are. So, for example, the most common word is "the," which we assign rank 1. The next most common word is "of," which is then given rank 2, followed by "and" having rank 3, "to" with rank 4, and so on. What Zipf discovered was that if we plot word frequency in a large body of English text versus word rank, we obtain a graph of the type shown in Figure 5.11.

Algebraically, Zipf represented this relationship between word frequency and word rank as an inverse one, in which the frequency de-

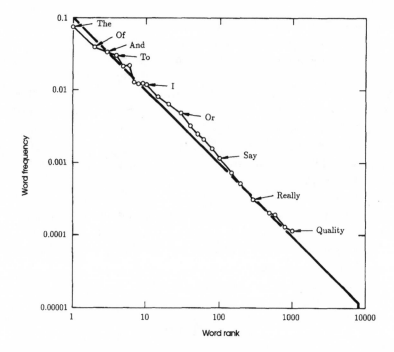

Figure 5.11. Zipf's law for the English language.

creased as the word rank became larger. Mathematically, he used the expression

$$f(r) \propto \frac{1}{r \log 1.78R},$$

where r is the word rank and R is the number of words in the language. Relations of the form $f(r) \propto r^{\alpha}$ are called power laws, in this case an *inverse power law* because here $\alpha = -1$. For a good writer with an active vocabulary of, say, $R = 100,000$ words, the ten highest-ranking words occupy 24 percent of a text, whereas for a popular novel or newspaper using a slimmed-down vocabulary of $R = 10,000$ words, this percentage increases marginally to nearly 30 percent. To illustrate an in-between case, William Bowers of Thousand Oaks, California, kindly sent me a Zipfian analysis of Conan Doyle's classic Sherlock Holmes story *The Hound of the Baskervilles.* Using a vocabulary size of $R = 10,000$ words, the result of Bowers's analysis of this 59,498-word story is shown in Figure 5.12. The striking conformance of the Conan Doyle story with Zipf's theoretical prediction is clear, although it is interesting to note that in this story the word "to," which holds rank 4 in general English text, has moved down to fifth place in the rankings for this particular text. Note also that in his analysis, Bowers has used the scale factor 1 rather than Zipf's original value of 1.78, a change of no consequence insofar as the inverse power-law relationship is concerned.

Zipf tried to derive the power-law form of his law by appealing to a principle of least effort. The appendix to this chapter shows how the derivation of the power-law structure of this law can be obtained by using arguments from information theory.

It turns out that monkeys hitting typewriter keys at random will also produce a "language" obeying Zipf's law. Suppose we consider a nine-letter alphabet with a space character, so that our mythical monkey strikes each character with likelihood 0.1. Many years ago, mathematician Benoit Mandelbrot showed that the exponent in Zipf's law is $\alpha = -1.048$ for this monkey language, hardly any different than the value of $\alpha = -1$ for English. But the median word rank in this language is 1,895,761, which means it takes this many of the most frequent words in the monkey language to reach a total probability of 0.5. By way of

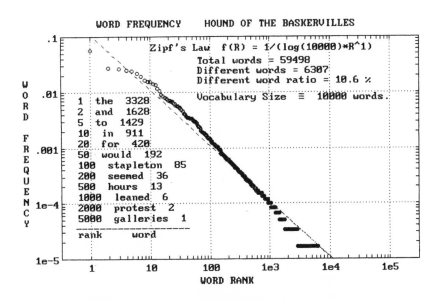

Figure 5.12. Zipf's law and *The Hound of the Baskervilles.*

contrast, the analogous figure in English is between 100 words (for typical media texts) and 500 (for belles-lettres writers). Thus, the monkey language is a very wordy one compared to English.

Somewhere in between the language of English and that of the monkey is the language of DNA. Recently, physicist Eugene Stanley, in collaboration with a group of geneticists, applied the Zipf test to the part of yeast DNA that does not correspond to the coding region for any genes, the so-called junk DNA. When the researchers arbitrarily divided up the junk into "words" between 3 and 8 bases long, what emerged was a surprisingly tight fit to the theoretical Zipf curve with exponent $\alpha = -1$, as shown in Figure 5.13. Moreover, the researchers applied a second test to quantify the "redundancy" in the yeast language, finding a level significantly higher than what would be expected if the "junk" were completely random. These two findings together suggest that something is written in these mysterious regions, and that the junk may not be so much junk, after all. As Harvard biologist Walter Gilbert described it, "I think the junk is like the stuff in a junk shop. You can find lovely things in it."

So what can we conclude from this consideration of Zipf's law? Is it really a "law" of languages encompassing everything from human to monkey to DNA? Or is it just an empirical curiosity, of no more system-theoretic significance than, say, a spurious correlation such as the Super Bowl indicator for the movement of stock prices? At this stage we would

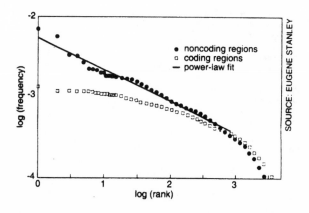

Figure 5.13. Zipf plot for arbitrary words in noncoding part of yeast DNA.

have to say that Zipf's law is still more of an empirical relation than a law. But the empirical evidence is mounting so that when we finally manage to create a theoretical structure within which to speak mathematically about complex systems, power laws such as Zipf's will appear as necessary *consequences* of such theories. When (and if) they do will be the moment when they pass from interesting empirical curiosities to laws of nature. When that day comes, information-theoretic reasoning will be shown to underpin the laws of complexity.

Appendix: Derivation of the Power Law Form of Zipf's Law

The argument for deriving the power-law form of Zipf's law is a bit cleaner, mathematically speaking, if we pretend that the word rank x is not a discrete quantity, but a continuous one. Then what we are searching for is the probability density function $p(x)$ characterizing a word of rank order x that maximizes the a priori entropy (uncertainty)

$$H(x) = -\int p(x) \log p(x)\, dx, \qquad \int p(x)\, dx = 1.$$

If we think about the maximum per-word information conveyed by a language using such a syntax, under a given constraint, we are led to the following functional of the unknown probability density function that we want to maximize:

$$S(p) = -\int p(x) \log p(x)\, dx - \lambda_1 \int p(x)\, dx - \lambda_2 \int \log x p(x)\, dx.$$

Here λ_1 and λ_2 are the Lagrange multipliers for this optimization. The density function p^* maximizing $S(p)$ turns out to be

$$p^*(x) \propto \exp\{-\lambda_1 - \lambda_2 \log x\} = \frac{A}{x^{\lambda_2}},$$

where A is a normalization constant over the interval of definition of the quantity x. This is exactly the form of Zipf's law.

Underlying this result is the natural assumption that in any language long words are used less frequently than short words. So the rank order of a word of length L, the quantity x_L, belonging to a language having an alphabet of K symbols ($K = 27$ for English, if we include the space) can be taken to be the sum of *all* the words of length not more than L. Thus

$$X_L = \sum_{i=0}^{L} K^i = \frac{K^{L+1} - 1}{K - 1}.$$

After a lot more algebra, which the reader can find in the chapter references, we end up with the relationship for L as

$$L \approx L_0 - \log_K L_0 + L' \log_2 X_L \approx L_0 + L' \log_2 X_L.$$

Therefore, the length of a word is approximately proportional to the logarithm of its rank order.

References

Note: Contrary to the normal custom, the entries in this list are not alphabetized. Rather, they are listed in approximately the order in which they apply to the flow of material treated in the text.

Chapter 1 (Knot Theory)

Adams, C. *The Knot Book.* New York: W. H. Freeman, 1994. A remarkable introduction to the theory of knots and their applications. First-rate exposition, coupled with just the right blend of figures and examples to bring out the intricacies of the field. Highly recommended.

Livingston, C. *Knot Theory.* Washington, DC: Mathematical Association of America, 1993. Marvelous exposition of the mathematical theory of knots. Easy enough for the motivated undergraduate to learn what's what with the theory; detailed enough for the professional to learn something. Marred only by a lack of material on the applications of knot theory in biology, physics, and chemistry. But you can't have everything, and this book gives plenty as it is.

Murasugi, K. *Knot Theory and Its Applications.* Boston: Birkhäuser, 1996. Graduate textbook on the topic. Distinguished by its careful attention not only to development of the mathematical theory, but also to a rather detailed attention to applications. (English translation of the 1993 Japanese original.)

Jones, V. F. R. "Knots in Mathematics and Physics," in *Symposium on the Current State and Prospects of Mathematics.* Berlin: Springer, 1992, 70–77. Expository account direct from the source of the relationship between knot theory and physics.

Rolfsen, D. *Knots and Links*. Berkeley, CA: Publish or Perish, Inc., 1976. Graduate-level set of lecture notes on geometric topology. Uses knots and links as a source of examples to illustrate techniques in algebra and topology.

Kauffman, L. *Knots and Physics,* 2d edition. Singapore: World Scientific, 1993. Essentially, a 300-page "short" course in knots with a strong emphasis on their use in physics. The second 300-page section of the book consists of a set of essays on miscellaneous topics at the interface between mathematics and physics, in which knots play a role.

Summers, de Witt. "Untangling DNA." *The Mathematical Intelligencer,* 12, no. 3 (1990), 71–80. Excellent exposition for the mathematically inclined of what knots and DNA have to say to each other.

Moffatt, H. K. "The Energy Spectrum of Knots and Links." *Nature,* 347 (27 September 1990), 367–369. Source article for the discussion in the text on the energy of a knot as a knot invariant.

Menasco, W., and L. Rudolph. "How Hard Is It to Untie a Knot?" *American Scientist,* 83 (January-February 1995), 38–49. Nontechnical introduction to knot theory for nonmathematicians.

Tabor, M., and I. Klapper. "The Dynamics of Knots and Curves–Part I." *Nonlinear Science Today,* 4, no. 1 (1994), 7–13. Expository account of the basics of knots and links for mathematicians. Nice introduction to the calculation of the Alexander polynomial.

Birman, J. "New Points of View in Knot Theory." *Bulletin of the American Mathematical Society,* 28, no. 2 (April 1993), 253–287. Technical overview of developments in knot theory that followed the discovery of the Jones polynomial. Emphasis is on insight into the topological meaning of the new invariants that followed from Jones's discovery.

Chapter 2 (Dynamical System Theory)

Hirsh, M., and S. Smale. *Differential Equations, Dynamical Systems, and Linear Algebra.* New York: Academic Press, 1974. Undergraduate textbook on dynamical systems written by two of the foremost researchers in the field. Highly recommended for its judicious blend of mathematical theory and interesting applications.

Guckenheimer, J., and P. Holmes. *Nonlinear Oscillations, Dynamical Systems, and Bifurcations of Vector Fields.* New York: Springer, 1983. First-rate graduate-level textbook on modern dynamical system theory.

Arnol'd, V. *Ordinary Differential Equations.* Cambridge, MA: MIT Press, 1973. Undergraduate textbook written by one of the masters of the dynamical systems art. Lots of geometric motivation, coupled with many

REFERENCES

examples from physics and mechanics, makes this an excellent book for either self-study or classroom use.

Jackson, E. Atlee. *Perspectives of Nonlinear Dynamics–1 and 2.* Cambridge, UK: Cambridge University Press, 1990. Comprehensive two-volume treatment of virtually every aspect of dynamical systems, yet surprisingly accessible to the nonmathematical reader. Emphasis is on the underlying concepts of dynamical systems rather than formal proofs. Highly illustrated with many examples, especially from physics. About as good an introduction to the mysteries and marvels of dynamical systems as you'll ever find.

Hahn, W. *Stability of Motion.* New York: Springer, 1967. Classical treatment of all aspects of the stability theory of differential equations. Written before the "chaos revolution," so the book focuses almost entirely on the classical attractors and the methods for their analysis.

Kaye, B. *Chaos and Complexity.* Weinheim, Germany: VCH, 1993. Introductory account of fractals presented using extensive examples from every aspect of life. Highly accessible even to those with no mathematical background at all.

Ott, E. *Chaos in Dynamical Systems.* Cambridge, UK: Cambridge University Press, 1993. Graduate text on modern chaos theory.

Laskar, J. "A Numerical Experiment on the Chaotic Behavior of the Solar System." *Nature,* 338 (16 March 1989), 237–238. Account of the author's numerical calculation of the largest Lyapunov exponent for planetary motion.

Carr, J. *Applications of Centre Manifold Theory.* New York: Springer, 1981. Detailed discussion of the Center Manifold Theorem, with many examples.

Casti, J. *Reality Rules–I.* New York: Wiley, 1992. Text-reference containing chapters outlining the theories of cellular automata, chaos, and catastrophe theory.

Mandelbrot, B. *The Fractal Geometry of Nature.* New York: Freeman, 1983. Standard reference for fractals and their applications. This is the book that sparked off the field.

Schroeder, M. *Fractals, Chaos, Power Laws.* New York: Freeman, 1991. Contains a wealth of examples of how chaos, fractals, and all the rest appear in everyday life.

Peitgen, H.-O., and D. Saupe, eds. *The Science of Fractal Images,* New York: Springer, 1988. Distinguished for its fantastic color graphics, making it possible to truly appreciate the endless complexity of things such as the Mandelbrot set.

Peterson, I. *The Mathematical Tourist.* New York: Freeman, 1988. Contains a good layman's account of the Mandelbrot and Julia sets, as well as a lot of other interesting modern mathematics.

Peitgen, H.-O., H. Jürgens, and D. Saupe. *Chaos and Fractals.* New York: Springer, 1992. Extensive account of the details of chaos and fractals. Filled with diagrams, games, computer experiments, biographical asides, color pictures, and just about every other kind of pedagogical trick to make the material accessible to the nonmathematically inclined.

Chapter 3 (Control Theory)

Casti, J. *Linear Dynamical Systems.* New York: Academic Press, 1987. Text-reference on all aspects of linear systems. Includes material on the algebraic-geometric aspects of linear system theory.

Fortmann, T., and K. Hitz. *An Introduction to Linear Control Systems.* New York: Dekker, 1977. Good undergraduate-level text on linear systems. Emphasis is on the engineering, as opposed to mathematical, aspects of the subject.

Brockett, R. *Finite-Dimensional Linear Systems.* New York: Wiley, 1970. Excellent combination of the mathematical and engineering aspects of linear systems. Easily accessible to undergraduates.

Casti, J. *Nonlinear System Theory.* New York: Academic Press, 1985. Graduate-level text on reachability, observability, and stability for nonlinear control systems.

"Recent Trends in System Theory." *Proceedings of the IEEE,* 64, no. 1 (1976). Special issue devoted to the modern theory of control systems. Of special note is the article by Roger Brockett on the uses of differential geometry for studying the problem of reachability for nonlinear systems.

Anderson, B. D. O., and J. Moore. *Linear Optimal Control.* Englewood Cliffs, NJ: Prentice-Hall, 1971. Extensive treatment of the linear-quadratic control problem.

Sontag, E. *Mathematical Control Theory.* New York: Springer, 1990. Graduate-level reference on mathematical system theory, with special emphasis on nonlinear processes.

Nijmeijer, H., and A. J. van der Schaft. *Nonlinear Dynamical Control Systems.* New York: Springer, 1990. Graduate textbook on nonlinear system theory. Notable for many exercises and worked examples.

Bellman, R., and S. Dreyfus. *Applied Dynamic Programming.* Princeton, NJ: Princeton University Press, 1962. Classic work on dynamic programming. Emphasizes applications in control theory and operations research.

Larson, R., and J. Casti. *Principles of Dynamic Programming–I and II*. New York: Dekker, 1978, 1982. Introduction to dynamic programming. Treatment is almost exclusively by hundreds of worked examples, emphasizing the numerical aspects of the subject.

Antoulas, A., ed. *Mathematical System Theory*. Berlin: Springer, 1993. Festschrift in honor of Rudolf Kalman. Contains essays covering all aspects of Kalman's many contributions to the field of mathematical system theory.

Grewal, M., and A. Andrews. *Kalman Filtering*. Englewood Cliffs, NJ: Prentice-Hall, 1993. Hands-on introduction to the theory and practice of Kalman filtering. Full of worked examples. Includes computer diskette containing many of the algorithms discussed in the text. Highly recommended for those who want to actually *use* the Kalman filter for practical problems.

Barnett, S. *Introduction to Mathematical Control Theory*. Oxford: Oxford University Press, 1975. Good undergraduate introduction to all aspects of the theory of control processes. Many exercises and no mathematics beyond linear algebra and calculus.

Chapter 4 (Functional Analysis)

Liusternik, L., and V. Sobelev. *Elements of Functional Analysis*. New York: Frederick Ungar Publishing, 1961. A classic introduction to functional analysis. Written in the Russian tradition of great expository writing, filled with many examples (from mathematics) as well as a thorough account of the theoretical concepts. Accessible to upper-division undergraduates.

Luenberger, D. *Optimization by Vector Space Methods*. New York: Wiley, 1969. Wonderful treatment of the methods and tools of functional analysis as they pertain to problems in optimization and control theory. Just the right blend of mathematical rigor and engineering practicality to show how functional analysis really works.

Zeidler, E. *Applied Functional Analysis—Main Principles and their Applications* and *Applications to Mathematical Physics*. New York: Springer, 1995. The first of this two-volume set gives a thorough technical treatment of all the important theoretical concepts in functional analysis, coupled with a lot of interesting worked-out examples, mostly from engineering and physics. The second volume focuses on the tools and spaces needed to deal with state-of-the-art problems in elementary particle physics, quantum theory, turbulence, and classical mechanics. All in all, an invaluable addition to every analyst's library.

Dieudonné, J. *History of Functional Analysis.* Amsterdam: North-Holland, 1981. Illuminating treatment of the development of functional analysis by one of the founders of the French Bourbaki movement. Gives an insider's glimpse of how ideas merge, separate, and intertwine as a field unfolds.

Mauldin, R. Daniel, ed. *The Scottish Book,* Boston: Birkhäuser, 1981. Translation of the book of problems from the Scottish Café in Lvov, together with personal recollections of the atmosphere of the period by Stanislaw Ulam, Mark Kac, Antoni Zygmund, Paul Erdös, and others. Wonderfully illuminating and full of insight into the life of the mathematical world in Europe prior to the outbreak of World War II.

Kaluża, R. *The Life of Stefan Banach.* Boston: Birkhäuser, 1996. In-depth life story by a reporter of the mathematician whose name is one of the most encountered in modern mathematical research. Presents an engaging description of Banach's personality and the very special milieu of mathematical life in Poland in the 1920s.

Goffman, C. "Preliminaries to Functional Analysis," in *Studies in Modern Analysis,* R. C. Buck, ed. Washington, DC: Mathematical Association of America, 1962. Undergraduate-level introduction to aspects of functional analysis, such as fixed-point theorems, Hilbert and Banach spaces, Banach algebras, and the Spectral Theorem. Also contains a short, but surprisingly complete, supplement on the historical development of the field.

Radjavi, H., and P. Rosenthal. "The Invariant Subspace Problem." *Mathematical Intelligencer,* 4, no. 1 (1982), 33–37. Excellent summary of the state of the Invariant Subspace Problem.

Packel, E. "Hilbert Space Operators and Quantum Mechanics." *American Mathematical Monthly* (October 1974), 863–873. Undergraduate exposition of how Hilbert space provides an elegant setting for mathematically describing the problems of quantum mechanics.

Van der Waerden, B. L. "From Matrix Mechanics and Wave Mechanics to Unified Quantum Mechanics." *Notices of the American Mathematical Society,* Vol. 44, No. 3 (March 1997), 323–328. First-hand historical account of how the matrix formulation of quantum phenomena due to Heisenberg and the wave function approach of Schrödinger were reconciled in the late 1920s. Great description of what the giants of modern physics— Heisenberg, Schrödinger, Pauli, Born, de Broglie—were doing to bring about the quantum revolution.

Sigmund, K. "A Philosopher's Mathematician: Hans Hahn and the Vienna Circle." *Mathematical Intelligencer,* 17, no. 4 (1995), 16–29. Very illuminating account of how the philosophers and mathematicians of

"Wittgenstein's Vienna" interacted in 1930s. The focus is upon the greatly underappreciated mathematician Hans Hahn, who founded the famed Vienna Circle of philosophy. Intertwining Hahn's life with that of Wittgenstein, Banach, von Neumann, Gödel, Popper and many others, this article is about the best introduction I know for mathematicians as to what was happening—philosophically and mathematically—in central Europe in this turbulent period.

Schwartz, J. *Nonlinear Functional Analysis.* New York: Gordon and Breach, 1969. Good introduction at a professional level of how functional analysis extends and generalizes calculus, the Implicit Function Theorem, degree theory, Morse theory, and a lot more. Hard going in places, but well worth the effort.

Collatz, L. *Functional Analysis and Numerical Analysis.* New York: Academic Press, 1966. A nice introduction to those aspects of functional analysis that have proven useful in numerical analysis.

Chapter 5 (Information Theory)

Shannon, C., and W. Weaver. *The Mathematical Theory of Communication.* Urbana: University of Illinois Press, 1964. Reprints of two papers on information theory. The first is a somewhat expanded version of an article that Weaver wrote for *Scientific American* magazine in July 1949. The second is a word-for-word reprint (with corrections of errata) of Shannon's famous 1948 papers in the *Bell System Technical Journal* that established the field of information theory. Indispensable reading for anyone interested in the subject.

Roman, S. *Introduction to Coding and Information Theory.* New York: Springer, 1997. Up-to-date introductory textbook on information and coding theory. Outstanding exposition of the basic ideas at the undergraduate level. Filled with many examples.

Losee, R. *The Science of Information.* San Diego: Academic Press, 1990. Exposition of "Information" with a capital I. Treats information in all its many forms, ranging from signals in the brain to language and on to the production and uses of information. Amazing survey in less than 300 pages.

Pierce, J. *An Introduction to Information Theory.* 2d, revised edition. New York: Dover, 1980. Another classic volume written primarily for the layman. Gives an account of information theory requiring very little background beyond high-school algebra, together with a discussion of applications ranging from physics to art. Marred a bit by its age, but still one of the best introductory sources, especially for some of the historical developments leading up to Shannon's work.

Nicolis, J. *Dynamics of Hierarchical Systems.* Berlin: Springer, 1986. Interesting exposition of several approaches to the modeling of two interacting, self-organizing hierarchical systems. Information theory section is a good, but slightly technical, introduction to the topic from a mostly engineering point of view.

Cover, T., and J. Thomas. *Elements of Information Theory.* New York: Wiley-Interscience, 1991. Graduate-level text on information and coding theory. Great problems. Includes a very interesting chapter on information theory and financial markets.

Yockey, H. *Information Theory and Molecular Biology.* Cambridge, UK: Cambridge University Press, 1992. First third of the book is devoted to an exposition of the mathematics of information theory. The remainder uses the mathematics to address a host of problems in biology, ranging from the origin of life to the genetic code. Thought-provoking.

Thompson, T. *From Error-Correcting Codes through Sphere Packings to Simple Groups.* Washington, DC: Mathematical Association of America, 1983. Extremely well-written introduction to the topics in the title. The opening sections on error-correcting codes are notable for their clarity of exposition as well as the detailed historical material surrounding Hamming's work.

Calude, C. *Information and Randomness.* Berlin: Springer, 1994. Extremely clear monograph on algorithmic information theory. Contains discussion of some interesting philosophical questions surrounding the meaning of "randomness," as well as notion of randomness in physics and the information content of mathematical knowledge.

Chaitin, G. *Information, Randomness, and Incompleteness.* 2d edition. Singapore: World Scientific, 1990. Collection of Chaitin's expository and technical papers tracing the development of algorithmic information theory.

Chaitin, G. *The Limits of Mathematics.* Singapore: Springer, 1998. Summary statement of Chaitin's life's work on information, randomness, complexity, algorithms, and mathematics. Basically, this is a course on the limits of formal reasoning. First half of the book consists of expository chapters on AIT, and the second half is an account of Chaitin's new version of *LISP* that enables one to implement in *real* computers his new definition of self-delimiting universal Turing machines. Heady stuff!

Li, M., and P. Vitányi. *An Introduction to Kolmogorov Complexity and Its Applications.* New York: Springer, 1993. Technical introduction to AIT. Final chapter on physics and computation is especially intriguing.

Index